The Evolutionary Road to Human Memory

The Evolutionary Road
to Human Memory

ELISABETH A. MURRAY
STEVEN P. WISE
MARY K. L. BALDWIN
KIM S. GRAHAM

OXFORD

UNIVERSITY PRESS

OXFORD
UNIVERSITY PRESS

Great Clarendon Street, Oxford, OX2 6DP,
United Kingdom

Oxford University Press is a department of the University of Oxford.
It furthers the University's objective of excellence in research, scholarship,
and education by publishing worldwide. Oxford is a registered trade mark of
Oxford University Press in the UK and in certain other countries

Excerpt from JURASSIC PARK: A NOVEL by Michael Crichton, copyright © 1990 by Michael Crichton.
Used by permission of Alfred A. Knopf, an imprint of the Knopf Doubleday Publishing Group, a division of
Penguin Random House LLC. All rights reserved.

Excerpt from THE HITCHHIKER'S GUIDE TO THE GALAXY by Douglas Adams,
copyright © 1989 by Pan Macmillan. Reproduced with permission of Pan Macmillan through PLSclear.

The moral rights of the authors have been asserted

First Edition published in 2020

Impression: 1

Published in the United States of America by Oxford University Press
198 Madison Avenue, New York, NY 10016, United States of America

British Library Cataloguing in Publication Data

Data available

Library of Congress Control Number: 2019947387

ISBN 978–0–19–882805–1

Printed and bound by
CPI Group (UK) Ltd, Croydon, CR0 4YY

To every child who ever loved a stegosaurus.

Preface

Apes and astronauts

In the novel *Planet of the Apes*,[1] a chimpanzee named Zira "ape-splains" evolution to a human astronaut. Marooned on the planet Soros, the astronaut prompts Zira by asking whether apes evolved from humans on her planet. "Some of us thought so," she begins, "but it is not exactly that."

> Apes and men are two separate branches that have evolved from a point in common but in different directions, the [apes] gradually developing to the stage of rational thought, the [humans] stagnating in their animal state. Many orangutans, however, still insist on denying this obvious fact.

To most of us, the astronaut's question seems topsy-turvy. How could apes have evolved from humans? Apes lack language and live in the open; humans talk ceaselessly and complain bitterly if their roof leaks—even a little. On the planet Soros, it's the other way around. Humans can't speak, and they live in the wild. Orangutans govern society; chimpanzees assist them (with more than a little resentment); and gorillas enforce the law. Orangutans also dominate the scientific establishment despite the fact that "a line of great thinkers, all of them chimpanzees" made all the big breakthroughs.

When the astronaut crash-lands on Soros, an injury renders him mute for a while, but that doesn't stop him from having something to say about the apes' view of humans as lower, "sub-ape" primates. Once his voice recovers, he intends to harangue an orangutan about this offensive attitude, but first he listens to Zira explain why her planet has garrulous gorillas and dumb humans. The "ape's brain," she claims, has become more "developed, . . . complex, and organized, whereas man's has hardly undergone any transformation."

> The ape's mind was primarily the result of the fact that he had four agile hands . . . With only two hands, each with short, clumsy fingers . . . man is probably handicapped at birth, incapable of progressing . . .

Clearly, Soros awaits the evolution of gender-neutral language. But despite her old-fashioned word choices, Zira gets a few things right. On Earth, at least, humans and apes did "evolve from a point in common."

But in her oration on origins, Zira gets a lot wrong. She claims that the human brain "stagnated" on Soros while the ape brain underwent Earth-shattering changes. As the novel unfolds, readers learn that Zira—a chauvinistic chimp—underestimates human intelligence. More importantly, her statement exposes a serious misconception about evolution. After two species split from their "point in common," both usually change. Although there are exceptions to this rule (called living fossils), neither humans nor apes were likely to have "stagnated" during evolution on Soros, and nothing of the sort happened on Earth either.

Zira also expresses frustration about getting her ideas accepted. "Many orangutans," she complains, "still insist on denying" obvious facts. Entrenched scientific doctrines are difficult to dislodge on planet Earth, too, and the result is the same as on Soros: enduring scientific error. In this book, as in our tome on the same topic,[2] we challenge a widely accepted theory about human memory. That view—the one you'll find on the internet, in textbooks, and in just about every brief summary of the subject—is simple and popular but fundamentally wrong. It assigns human memory to a small corner of the brain called the "medial temporal lobe." This one brain area is said to be responsible for all of the memories that shape the human mind.

So, what's our beef with this idea? To begin with, the medial temporal lobe makes up about 2% of the brain, and it seems strange that such a tiny bit of brain would underlie all human memory. What's the rest of the brain doing? This question is simple, but our answer is not. We believe that as a series of now-extinct ancestors adapted to life in their world, millions of years ago, new brain areas appeared, each of which supported an innovative form of memory that we've inherited in modified form. Taken together, these areas make up most of the brain: closer to 80% of it than to 2%. Memory, at least as we see it, reflects a journey along a long evolutionary road, which left its legacy in the human brain.

Animals and attitudes

In Chapters 9 and 10, we'll discuss some ways in which human memories differ from those in other animals. It's a hard sell, in part because claims of human exceptionalism have a checkered past. Scarcely a week goes by without a news article proclaiming that some species has a cognitive capacity previously thought to be uniquely human. Indeed, so many prior claims have ended up on the scrapheap that one scientist, Frans de Waal, has called for a moratorium on claims of human uniqueness. Similarities between human and parrot vocalizations are said to discredit the stodgy doctrine that only people can talk; claims about ape "language" attract endless publicity; and many animals are said to have the same tool use, reasoning, and emotional experiences as people do. These articles imply that animals and humans have the same cognitive capabilities, but narrow-minded, pig-headed scientists like us refuse to accept this enlightened attitude.

Despite objections to human exceptionalism, many scientists continue to recognize a substantial cognitive divide between humans and animals.[3-6] Slowly, the idea that each species has its own distinctive set of cognitive capacities is gaining ground. We mean no offense to other species when we say that their cognitive lives differ from ours. Three of the authors have spent their entire careers studying animal brains and behavior. Collectively, and in no particular order, we've studied rats, cats, and bats; bears and bushbabies; tree shrews and turtles; squirrels, opossums, chickens, voles, mice, and ferrets; along with six species of monkeys. We admire these and other animals for what they are: not exactly like us, but not completely different either. That statement isn't an opinion; it's a fact, one that results from another, equally well-established fact: humans and every animal species alive today have an equally long evolutionary history, much of it common but recently separate. Both parts of our evolutionary history—the common part and the separate part—have left their mark on human memory.

Areas avoided

We neglect several topics that you might expect to find in a book on memory. Birds, for instance, have astonishing memories,[7] often surpassing anything we can do. Songbirds establish exquisite memories of their songs, which they never forget; people are known to bungle the lines of their national anthem. Scrub jays have precise memories about where and when they stashed food; some of us have trouble remembering where we've put our keys. But we don't have any birds, or anything like a bird, among our ancestors. Although we share many extinct ancestors with birds, our last common ancestor neither sang like modern songbirds nor cached food like jays do today. We are vertebrates, mammals, primates, and anthropoids, among other things; but we are not birds, so we don't discuss them much in this book.

We also avoid ethics, religion, and philosophy. Although we mention consciousness from time to time, we ignore its neural basis for the most part. We don't deal with creationism or the pseudoscience called "intelligent design." And, because we know that animal research is necessary for a deep understanding of the brain and its disorders, we rely heavily on this crucial branch of science and ignore any controversies about it.

Finally, we say almost nothing about how the brain establishes memories, how the brain develops and matures, how to improve memory, or how to deal with memory disorders such as Alzheimer's disease. Nor do we provide a comprehensive description of human memory. These are all important topics, but not for the story that we want to tell. And the gist of that story is this: as a series of our direct ancestors faced the problems and opportunities of their time and place, their brains developed new forms of memory that helped them gain an advantage in life. Sometime during human evolution, yet another new kind of memory emerged. It ignited the human imagination, established ownership of knowledge, and empowered every individual, day upon day, to add new pages to the story of a life.

References

1. Boulle, P. *Planet of the Apes* (New York, NY: Ballentine, 1963).
2. Murray, E. A., Wise, S. P., & Graham, K. S. *The Evolution of Memory Systems: Ancestors, Anatomy, and Adaptations* (Oxford, UK: Oxford University Press, 2017).
3. Penn, D. C., Holyoak, K. J., & Povinelli, D. J. Darwin's mistake: explaining the discontinuity between human and nonhuman minds. *Behavioral and Brain Sciences* 31, 109–130 (2008).
4. Passingham, R. E. *What Is Special about the Human Brain?* (Oxford, UK: Oxford University Press, 2008).
5. Suddendorf, T. *The Gap: The Science of What Separates Us from Other Animals* (New York, NY: Basic Books, 2013).
6. LeDoux, J. E. & Pine, D. S. Using neuroscience to help understand fear and anxiety: a two-system framework. *American Journal of Psychiatry* 173, 1083–1093 (2016).
7. Emery, N. J. *Bird Brains: An Exploration of Avian Intelligence* (Princeton, NJ: Princeton University Press, 2016).

Acknowledgments

We thank Brad Postle and Georg Striedter for helpful comments on a draft version of this book, along with Peter Rudebeck and Carly Jones for comments on selected chapters. Samantha White and Hannah Goldbach gave us invaluable feedback over the summer of 2017, and we thank Mark Laubach for arranging those discussions. We are also grateful to our editors at Oxford University Press, Martin Baum and Charlotte Holloway, who provided support throughout this project.

Contents

It's a Long Way from Amphioxus[*]

. . . My tiny dorsal nervous cord shall be a mighty brain
And the vertebrates shall dominate the animal domain.

It's a long way from *Amphioxus*; it's a long way to us.
It's a long way from *Amphioxus* to the meanest human cuss.
So, goodbye fins and gill slits, welcome lungs and hair!
It's a long, long way from *Amphioxus*, but we all came from there.

[*] Sung to the tune of "It's a Long Way to Tipperary," one version of the lyrics is available at https://bscd.uchicago.edu/content/long-way-amphioxus, and a recording of the song by Sam Hinton can be purchased for $0.99 as Track 209 at https://folkways.si.edu/sam-hinton/sings-the-song-of-men/american-folk/music/album/smithsonian

1

A drive down memory lane

Imagine driving down a steep, one-lane mountain road in dense fog, unable to turn around, unwilling to stay put, and incapable of seeing the way forward. Our evolutionary ancestors faced something similar. As they traveled through their lives, they could neither turn back nor see into the future. Instead, they put everything they had into staying in their lane, surviving in the moment. From our vantage point, we can look back and appreciate how we got here, knowing three things for sure: only one path led to us; routes equally long led to all the species that inhabit our planet; and some trails led their travelers off a cliff. In some ways, evolution resembles a long drive along a mountain road, shrouded in a dense fog. What could possibly go wrong?

Roads pose dangers enough in real life but even more as a metaphor for evolution. Although travel usually implies progress toward a goal, evolution doesn't have any such thing. Instead, it's a blind, selective process with winners and losers, to be sure, but it doesn't aim to go anywhere. From time to time in the chapters to come, we'll refer to *The Wizard of Oz*. We do so in part because it's a tale about a treacherous journey and in part because it draws on shared cultural knowledge: two of our main themes. Unlike evolution, however, Dorothy and her companions have a definite aim in mind. As everyone knows, they intend to follow the yellow brick road from Munchkinland to the Emerald City. To understand the evolution of memory, we need to do things backward. It's a little like starting in the Emerald City and tracing the yellow brick road back to Munchkinland, then imagining an aimless journey starting out from there.

Roads also serve as a metaphor for memory science. Just as our animal ancestors motored along their evolutionary byways, memory research has traveled some precarious paths of its own. Like the real thing, some research roads are hard to follow, make sharp and unexpected turns, and can head in the wrong direction at times. Enveloping fogs develop quite frequently, and promising paths fall into disrepair. Unlike evolution, however, we can choose the course of memory research. By taking our goal into account—understanding human memory—we can turn around, head back, and explore a different direction if need be. In essence, we want to do something like that in this book. Later in this chapter,* we'll explain a particular choice made by memory scientists in the 1980s and 1990s and why we want to drive in a different direction.

---

* In the section entitled "A fork in the road."

The Evolutionary Road to Human Memory. Elisabeth A. Murray, Steven P. Wise, Mary K. L. Baldwin, and Kim S. Graham, Oxford University Press (2020). © Oxford University Press.
DOI: 10.1093/oso/9780198828051.001.0001

Memories and mindsets

This book tackles three topics—the cerebral cortex, memory, and evolution—and how they relate to each other. The cortex dominates our brains; memories guide our thoughts; and both have a long evolutionary history. Yet evolution has rarely influenced the science of human memory, and cortical evolution is often misunderstood—even by expert neuroscientists. Consequently, a simple idea about the cerebral cortex has reigned for more than a century: that it's organized into areas that specialize in either memory, sensory perception, the control of body movements, or something called executive control, which amounts to planning and organizing behavior. According to this view, only a few cortical areas specialize in memory; the vast majority do something else.

We advocate a different idea, one that takes evolution into account. In our opinion, every cortical area contributes to memory, each in a specialized way. As our ancestors traveled along their evolutionary trajectories, cortical areas accumulated over time; and, in each instance, this happened for the same fundamental reason: to transcend problems and exploit opportunities that these animals faced in their time and place. The innovations of our ancestors—as modified over millions of years of evolution—influence human memory to this day.

The first patient

People have pondered their memories from time immemorial, but contemporary ideas on the subject stem from a man named Henry Molaison, who suffered from epilepsy. Like other patients with this brain disorder, he experienced periodic seizures. His story has been told many times, most comprehensively by Suzanne Corkin,* who knew him for more than 50 years. Until his death, most scientists knew only his initials, H. M., but we'll call him Henry. We mean no disrespect, but "Henry" seems appropriate, not only because one of the authors knew him personally, but also because it lets us tell his story in a way we think he might have liked.

Henry had a severe and untreatable form of epilepsy, and in his early years he suffered several seizures every day. In 1953, a neurosurgeon removed the brain areas that caused his condition. As intended, the frequency of Henry's seizures decreased after the surgery, but the procedure also had an unfortunate consequence. Henry suffered a severe memory loss, which scientists call *anterograde amnesia*, a clinical term we'll explain shortly. After his surgery and for the rest of his life,

---

* We'll recommend further reading in footnotes like this one. Corkin's book, *Permanent Present Tense* (Basic Books, New York, NY, 2013), relates Henry's life and condition in detail. Superscripted endnotes provide documentation for important points.

Henry couldn't remember new acquaintances, recently encountered knowledge, or stories he had heard or read lately.

In the decades that followed, scientists noted a similarity between Henry's condition and that of other patients. These individuals also experienced a severe memory loss, some after surgery like Henry, but also as a consequence of progressive brain disorders, strokes, oxygen deprivation, or viral infections of the brain. Taken together with Henry's case, the study of these patients drove memory research along two roads. The first one focused on identifying the kinds of memory lost in anterograde amnesia; the second explored the brain areas needed for normal memory. In this chapter, we trace both of these roads and introduce a third one, which our ancestors have followed for more than 500 million years.

The first road

Since the earliest descriptions of Henry's condition, scientists have thought about anterograde amnesia in terms of memory systems. The reason is simple: Henry lost some forms of memory almost entirely, while others remained relatively normal. These observations seemed to imply that the brain must have separate memory systems: one for the kinds of memory disabled by Henry's surgery; others for memories he retained. Memory scientists have long assumed that Henry's lost memories depended upon the brain areas removed during his surgery and that his normal memories relied on other parts of the brain.

Memories labeled

The kinds of memory that Henry lost go by two roughly equivalent names: *declarative memory* and *explicit memory*. The first refers to the fact that people can tell each other about such memories; the second highlights the impression that these memories are "out in the open" and accessible. When most people use the word memory, they usually have the declarative, explicit form in mind. In this book, we call this kind of memory *personal*. This label emphasizes the idea that our memories belong to ourselves and no one else. *We* know the deadline for finishing this book; *we* know that Aunt Tex flew airplanes for the military during World War II; and *we* know that we watched *The Wizard of Oz* last night. Our memories are *personal* to us.

Experts commonly recognize two kinds of personal memories: *episodic* and *semantic*. We don't endorse this dichotomy, but it's important to understand these terms because they appear in many discussions of memory. Your *episodic memories* record what it was like when you experienced specific events in daily life. When recollected later, they enable you to "relive" an experience (at least to an extent); and when collected over a lifetime they compose your mental autobiography: the story of your life. A sense of participation in events is a crucial part of these memories, both as events occur and later, when you remember your experiences or share them with others. Accordingly, we use the term *participatory* for such memories.

The term *semantic memory* refers to your knowledge about the world, including the meaning of words and other symbols. For example, when a past president of the United States denied having sex with a White House intern, he was making a semantic argument. His claim was that the word "sex" didn't apply to their behavior, although many people disagreed.* In addition to words and symbols, semantic memory includes knowledge about categories and concepts, information that facilitates generalizations about the world and the things in it. For example, even if you've never met a chimpanzee up close and personal, you know that you can't have a conversation with one, except on the imaginary *Planet of the Apes*.

Experts often refer to semantic memory as "fact memory," but this label raises some tricky issues. It is fashionable these days to talk about a post-fact world and "alternative facts," but people have always believed in fictions. Young Earth creationists, for example, think that our planet is only 10 000 years old: an estimate that's wrong by a factor of 450 000 or so. In addition to falsehoods masquerading as facts, our cultural life includes a wealth of acknowledged fictions: short stories, novels, and plays among them. Because semantic memory includes falsehoods, fictions, categories, concepts, inferences, and generalizations, we'll call it *cultural memory* in this book rather than either fact memory or semantic memory.

It's often useful to distinguish between memories of facts and memories of events, but these classifications fail to capture the more nuanced distinctions among memories. We learn many facts through participation in informative events, which often remain in memory as both facts and events; and events are also facts (about what happened). Some facts come from our direct personal experience in real-world events, but most come from indirect sources: mainly Facebook®, Wikipedia®, and YouTube®, of course, but on rare occasions teachers, poets, and parents, too. What's more, some memories about our own experiences, which would typically be classified as event memories, come from exposure to family photographs or videos. These records provide information about events, but without many of the observations that we would have and remember from direct experience.

We hope that our new labels will help readers understand our main themes, but we recognize their downside. The familiar labels, the ones regularly used in memory research, help scientists understand each other. To ease the burden on such readers, from time to time we remind them what our new terms mean with phrases like personal (explicit) memory, participatory (episodic) memory, and

---

* He also quibbled about "what the meaning of the word 'is' is," perhaps the most obscure semantic argument in history.

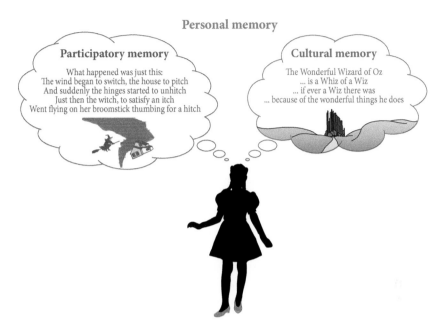

Fig. 1.1 Human memory. Together, participatory memory and cultural memory compose what we call personal memory. In the specialty literature, these kinds of memory are called episodic, semantic, and explicit (or declarative), respectively. Dorothy remembers a particularly striking event in the left thought-bubble, in the sepia tones of Kansas. It seems that a twister picked up her farmhouse, illustrated to the right of the storm, with the Wicked Witch of the West looming to the left. In the right thought-bubble, Dorothy considers some generalized knowledge about the Wonderful Wizard of Oz, in full color. Like a lot of cultural knowledge, some of the things Dorothy thinks she knows about the wizard aren't 100% accurate.

cultural (semantic) memory. In Figure 1.1, Dorothy considers our new labels in light of her experiences and knowledge.

Unfortunately, a poor choice of labels can impede scientific progress. For example, the use of a noun instead of a verb for *fire* is said to have retarded the understanding of fire as a dynamic physical and chemical process, as opposed to a static "thing." A similar problem occurred in memory science. Experts began referring to Henry's memory loss as "global amnesia." The use of the term "global" implies that his amnesia involved every kind of sensation—sights, sounds, smells, tastes, and touch—and some scientists have extended this concept to cover every kind of personal memory. One of the crucial discoveries of memory research, which we'll explain in Chapter 7,* is that the memory loss in amnesic patients is nowhere

* In the section entitled "Bugs and blobs for brains."

near global. It doesn't even include all forms of visual memory. Instead, it involves the loss of specific kinds of memory and the preservation of others, leading to far-reaching—but far from global—consequences.

Memories lost and retained

In Henry's case, his surgery caused a selective inability to establish new personal memories. At the same time, he could remember quite a bit from the years before his surgery, and these old memories enabled him to conduct a wide range of activities. He could converse with people using remembered word meanings, appreciate the fairness or unfairness of offers made to him, and solve mathematical problems that depended on principles he'd learned in school. You could chat with Henry and never suspect that he had a serious memory impairment.

Even after his surgery, when he had severe amnesia, Henry could hold onto new personal memories briefly. It is common to distinguish between memories maintained in mind for a few seconds or minutes, called *short-term memory* or *working memory*, and memories that can be retrieved after hours, days, or years, called *long-term memory*. As long as nothing distracted him, Henry could remember a three-digit number for about as long as most people can. He could also manipulate information in memory, such as adding or subtracting numbers he had just heard. So, Henry's specific problem was that he couldn't establish *new, long-term, personal* memories. That's the impairment that scientists call anterograde amnesia.

Before setting Henry's amnesia aside for a while, one final point: a clinical term like anterograde amnesia strikes us as too cut-and-dry, and it provides little appreciation of his life after surgery. Henry could never again live independently, learn anything about the world, or develop long-term relationships with other people. As he once said,[1] his life was like "waking from a dream . . . every day is alone in itself." That statement reveals something that clinical labels cannot convey. Every day, you add something to the story of your life as you experience it. Your memories provide an up-to-date record of your family, friends, achievements, and goals, and they empower you to transcend immediate concerns. After his surgery, Henry couldn't do that anymore.

The second road

Since the 1950s, memory scientists have known that Henry's neurosurgeon removed several parts of his brain, and they wanted to know which ones mattered most for memory. At first, they focused on an unusual part of the cerebral cortex called the hippocampus. (It's so unusual that even expert neuroscientists sometimes forget that it's part of the cerebral cortex.) Its name comes from a curvature that reminded early brain anatomists of a seahorse: *hippocampus* in Greek. Memory researchers first attributed Henry's amnesia to the removal of this brain area. For example, a landmark 1957 article was entitled "Loss of recent memory after bilateral hippocampal lesions,"[2] and two years later another one was called "The memory defect in bilateral hippocampal lesions."[3] As one textbook from 1970 put it[4]: "there is little doubt that the hippocampus is involved in the memorization of new materials."

With these ideas in mind, scientists attempted to replicate Henry's amnesia in monkeys, and they began by removing the hippocampus. Unfortunately, despite many attempts from the late 1950s until the mid-1970s, these experiments invariably ended in failure and frustration. In most memory tests of that era, monkeys faced a choice between two objects. If they chose the correct one, they got some food. By remembering which object had the most value, through its association with food, they could pass the test. As it turns out, the ability to choose an object based on a memory of its value doesn't depend on the hippocampus. But scientists didn't know that at the time. They only knew that nothing they did to the hippocampus could block or dislodge such memories. How could this be possible, given Henry's striking loss of memory?

To answer this question, some scientists of the time invoked the idea of species differences, a vague notion implying that although the hippocampus might be at the heart of personal memory in people, it has a different function in monkeys. Few memory researchers share that opinion today, and one reason is that interest in species differences went out of fashion in the 1980s and 1990s, at least for memory research. By that time, most experts had come to the opinion that monkeys and humans have the same kinds of memory, which rely on the same brain areas. In time, this idea expanded to include all mammals[5] and perhaps all vertebrates.[6] In this book and in our previous one,[7] we revisit research on species differences and show how it can inform our understanding of both the hippocampus and human memory. We think that the hippocampus *does* have some different functions in humans, compared to monkeys, as we'll explain in Chapter 10.

Although memory scientists from the 1950s onward mentioned species differences now and then, they didn't really embrace the idea. Mostly, they assumed that animals—and especially monkeys—serve as "models" of humans, and so they were perplexed by their failure to mimic human amnesia by removing or disabling the hippocampus in monkeys.

The matching test

Then, after decades of disappointment, at a time when it seemed there was nothing scientists could possibly do that would ever cause amnesia in monkeys, at long last something did. In 1974, David Gaffan broke the long string of failures.[8] To perform his breakthrough experiment, Gaffan assembled several dozen objects that differed from each other in color, shape, size, and other visual features. His monkeys could tell them apart easily. Each trial of his memory test began when a monkey saw one of the objects. After that particular object had been out of sight for a while, the monkey saw it again, along with a second object. The monkey had already learned that food would be hidden under the object it had seen at the start of the trial. So, the monkey simply needed to remember and choose that object to obtain some food. To tax memory, Gaffan increased the amount of time that the original object remained out of sight, a period called a *memory interval* or *delay period.* If the monkey wanted a snack—as monkeys nearly always do—it had to remember the first object for the entire delay period. Eventually, this test and its variants came to dominate memory research on animals. It has a jaw-breaking name in the specialty literature—the delayed matching-to-sample task—but we'll call it simply the *matching test.*

In a group of test monkeys, Gaffan disrupted the function of the hippocampus by cutting some of its connections with other parts of the brain. These monkeys passed the matching test when the memory interval lasted only 10 seconds. This observation was important because it showed that, despite their brain surgery, the monkeys still wanted a snack and could see the objects, tell them apart, and remember the rules of the test. As Gaffan lengthened the memory interval by a minute or two, his monkeys began to fail the test. When they had to remember the original object for slightly more than two minutes, their performance deteriorated to little better than they could achieve simply by guessing: 50% correct choices. In contrast, a group of normal monkeys got well over 90% of their choices correct at all delay intervals.

A few years later, in 1978, Mort Mishkin performed a similar experiment.[9] Instead of disrupting the function of the hippocampus by cutting its connections, as Gaffan had done, Mishkin removed the hippocampus to better emulate Henry's surgery. Mishkin's procedure also resembled Henry's surgery in another way; it involved removal of the *amygdala* as well as the hippocampus. *Amygdala* is the Latin word for almond, and its name reflects the shape of this brain structure. The brain's amygdala is much more complex than an almond, however; it consists of a diverse collection of structures often associated with emotion and motivation. In his experiment, Mishkin removed the hippocampus from one group of monkeys, the amygdala from another group, and both the hippocampus and the amygdala from a third group. In Figure 1.2d, we illustrate the locations of both brain areas in the human brain, and this aspect of brain anatomy is similar in monkeys. The first and second groups of monkeys passed the matching test, but the third group—the one with combined removals of the hippocampus and the amygdala—failed miserably. From these results, Mishkin concluded that a combined removal of the amygdala and the hippocampus caused amnesia in his monkeys and, by extension, in Henry.

Then, in the 1990s, a team of scientists headed by the first author of this book carried out a different experiment while working in Mishkin's laboratory. When Betsy Murray and her colleagues removed two cortical areas neighboring the amygdala and hippocampus—called the *entorhinal cortex* and the *perirhinal*

(a)   View from the left (lateral view)   (b)   View from the middle (medial view)

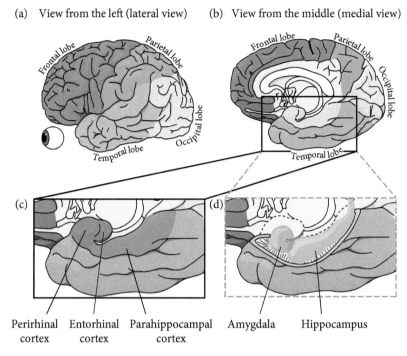

Fig. 1.2 The human brain. (a) The four lobes of the cerebral cortex, as viewed from the left side with the forehead facing left. (b) The right side of the brain with the left half removed. This view shows the middle (medial) parts of the cerebral cortex, where the two hemispheres face each other. The rectangle shows a part of the temporal lobe depicted, in different ways, in parts c and d. (c) The locations of the perirhinal cortex, the entorhinal cortex, and the parahippocampal cortex. (d) The locations of the amygdala and the hippocampus illustrated as if the medial surface cortex has been removed. The perirhinal cortex and the entorhinal cortex cover the amygdala when viewed from this perspective.

*cortex**—monkeys failed the matching test.[10] As we illustrate in Figure 1.2c and d for the human brain, these two brain areas, along with another one called the *parahippocampal cortex*, surround the amygdala and the hippocampus from the underside surface of the brain, and this configuration is similar in monkeys. A follow-up experiment involved removal of the amygdala and the hippocampus by injecting a toxin into these two brain areas.[11] Although this procedure took

---

  * The names *perirhinal* and *entorhinal* have nothing to do with the function of these cortical areas, despite the fact that *rhinal* refers to the nose or the sense of smell. Their names instead indicate their location around (*peri*) and within (*ento*) a small grove in the cortex called the *rhinal sulcus*. The name *parahippocampal* means that this area lies alongside the hippocampus, and not that it is any more like the hippocampus than any other cortical area.

more time and effort than the surgical removal that Mishkin had performed in 1978, it had the advantage of avoiding damage to nearby brain areas and connections. After this procedure, monkeys passed the matching test.

Murray's findings revealed that monkeys failed the matching test because of damage to their entorhinal cortex and perirhinal cortex, not because of a combined removal of the amygdala and the hippocampus. Mishkin had reached the wrong conclusion for two reasons. First, he removed the covering cortex that blocked direct access to the amygdala and the hippocampus. For some reason, he didn't think that these areas were important for passing the matching test, but they were. Second, he inadvertently cut connections between the perirhinal cortex and other brain areas. Cutting connections is like cutting a cable that goes to a video screen; it blocks a function—in this example, a video stream on the screen—although neither the source of the signal nor the video monitor suffers direct damage. Inadvertent damage to connections going to and from the perirhinal cortex eliminated its ability to help monkeys pass the matching test.

Although Murray's findings overturned Mishkin's original conclusions, most scientists accepted them for more than 20 years. This era in memory research had three unfortunate consequences, all of which persisted for decades and some of which continue to plague the field today:

- As we mentioned earlier, the idea of species differences virtually vanished from memory research. Monkeys were said to have the same kinds of memories as people do,[12–15] an assumption that scientists eventually extended to rats, mice, and other mammals.[5] A couple of scientists have even stretched this idea to cover all vertebrates.[6]
- Scientists came to depend almost exclusively on the matching test as a measure of personal (explicit) memory in monkeys.
- Because Mishkin concluded that the ability to pass the matching test depended on the amygdala as well as on the hippocampus, memory researchers turned their attention away from the hippocampus, alone, and toward a larger set of brain areas they called the "medial temporal lobe memory system." The collection of brain areas included under this umbrella term changed over time, but in Mishkin's original theory it consisted of the amygdala and the hippocampus.

A fork in the road

By the time Murray's results emerged, the idea of a "medial temporal lobe memory system" had become entrenched in the minds of memory scientists. Her evidence proved, however, that monkeys didn't need their hippocampus to pass the matching test. If this task truly assesses personal memory, as assumed, her results would have meant that the hippocampus plays no role in this form of memory in monkeys. But the hippocampus clearly does in humans, so in order to keep the hippocampus in the so-called medial temporal lobe memory system of monkeys, scientists faced a fork in the road. Down one path, they could abandon the idea that the matching test assesses personal memory in monkeys; down the other, they could enlarge the concept of the "medial temporal lobe memory system" to include the cortical areas that monkeys need to pass this test.

The first choice seems preferable today. For one thing, people can pass the matching test without using their personal memories. For another, amnesic patients like Henry can remember things over the short time intervals usually used in this test: a few minutes or less. So, the matching test doesn't seem like a good way to assess long-term, personal memories in monkeys or humans. Had the scientists of the 1980s and 1990s ditched the matching test, the hippocampus might have regained its pre-eminent place in the study of personal memory, as assumed in the 1950s. Equally important, they might have focused on interactions between the hippocampus and brain areas beyond the "medial temporal lobe memory system"; they might have reconsidered the possibility of species differences; and they might have considered the many ways in which human brains differ from those of monkeys. Unfortunately, none of that happened.

Instead, the leading memory scientists of the 1980s and 1990s made the other choice. A series of experiments conducted in the laboratories of Larry Squire and Mort Mishkin[12–15] led to the view of memory that dominates the field today. These researchers decided to retain both the concept of a "medial temporal lobe memory system" and the assumption that the matching test measures personal (explicit) memory in monkeys. Without much fanfare, however, they replaced the amygdala with three nearby cortical areas: the perirhinal cortex, entorhinal cortex, and parahippocampal cortex (in Fig. 1.2c, we depict these three cortical areas in the human brain). According to Squire, Mishkin, and their colleagues, personal memory depends on these three cortical areas, along with the hippocampus, in both monkeys and humans. They also settled on the idea that damage to these areas results in a "global" amnesia that spares perception: the ability to sense and distinguish things. Perception, according to these authorities, depends on parts of the cortex outside the "medial temporal lobe memory system."

This set of ideas has dominated memory science for more than 30 years, and at its heart is a denial of species differences. Monkeys, humans, and other mammals, we are told, have the same kinds of memory and the same "memory areas" in the brain.

A third road

A third road toward understanding human memory, the one traveled in this book and its predecessor,[7] depends entirely on species differences. It leads to what we call the *evolutionary accretion model* of memory, which has four main tenets: (1) memory comes in many forms, each of which depends on cortical specializations that evolved in a particular ancestral species; (2) every cortical area contributes to memory, each according to its specializations; (3) when they first evolved, each specialization provided a selective advantage over pre-existing ones; and (4) each species has its own combination of specializations. After a brief summary of brain evolution in Chapter 2, we'll explain these ideas in Chapters 3–10.

Although we avoid technical terms when we can, two concepts—*representation* and *homology*—underpin most of what we say. So, it's important to spell out what they mean.

Representation

When neuroscientists say that the brain contains a *representation* of pigs, for example, they mean that some information—about a particular pig, about pigs in general, about the abstract idea of a pig, or about something that resembles pigs in some way—is processed and stored by an interconnected network of neurons. Neural networks can be confined to a single brain area, but representations might also emerge from neurons connected across different areas.

All cognition—a word that refers to what we know and think about the world—depends on representations in the brain. Neural representations not only underlie our memories, but also our ability to perceive the world and control our behavior. They depend on two properties of neurons. First, neurons generate electrical pulses, often called spikes or discharges, that travel along axons to their ends, known as terminals. Importantly, the frequency and pattern of pulses change over time, often rapidly. Second, the terminals from each neuron form synapses with other neurons. Synapses transmit signals from neuron to neuron, and the strength of each synapse can change with experience. The brain holds information in both pulses and synapses, with the former called *neuronal activity* and the latter called *synaptic weights*. The neuronal-activity form of memory is like a video stream; it provides information as long as the stream persists. The synaptic-weight form is more like a downloaded video file, stored for future viewing.

Neuronal activity and synaptic weights explain *how* the brain establishes representations but not what neural representations *are*. Take, for example, the combination of contours and colors that compose the ruby slippers in Figure 1.1. Somewhere in your brain, an interconnected network of neurons represents ruby slippers as this combination of sensory features. When you see Figure 1.1, sensory inputs activate this neural network. Other neural networks can augment these sensory representations by combining them with additional information, such as the fact that you can see a pair of the original ruby slippers in the Smithsonian Institution.* Neural networks can also combine sensory representations with information about their value: how much you like or want the ruby slippers.

In discussions of animal and human cognition, the concept of neural representations—with its inherent focus on what happens inside the brain—is sometimes contrasted with the idea that cognition depends on interactions with the outside world. But these ideas don't conflict in any way. The brain controls the body and its actions; actions affect the local environment; and the local environment affects the body as well as the brain. So neural representations depend on the continual interplay between an individual and its surroundings. In the specialty literature, scientists and philosophers discuss these ideas in terms of *embodied cognition* and *enacted cognition*. We avoid this terminology but embrace the attitude it embodies (no pun intended). The brain is part of an extended feedback system in which behavior changes the sensations that an animal experiences[16]; and sensations often guide behavior.

Homology

This book also depends on the concept of homology, which refers to heritable traits passed down from ancestors to descendant species. The term *trait* applies to any characteristic of a species, especially distinctive ones. Traits may be physical, like the snouts and hooves of pigs, or behavioral, like their rooting and grunting.

The basic idea is that a given species can sometimes split into two groups. As time goes by, these two populations of individuals interact with different environments, and natural selection acts differently on their heritable traits. Eventually traits change, typically in both populations, as they become two different species.† In this way, the forelimbs that helped move a common ancestor along the ground can morph into both forelimb flippers that propel sea turtles through water and wings that lift parrots into the air (both of which we illustrate in Fig. 1.3b).

In Figure 1.3, we present the kind of diagram that depicts evolutionary relationships. Think of these drawings as upside-down trees, with time running from top to bottom. At the top, the tree's trunk represents the last common ancestor, and branches divide toward the bottom, closer to modern times. Each line depicts a lineage of direct descent from the last common ancestor to modern or extinct species;

---

* See http://americanhistory.si.edu/exhibitions/ruby-slippers-and-american-culture-displays
† This book is about memory, not evolution, so we won't say any more on this topic. A recent book by Jonathan B. Losos, *Improbable Destinies: Fate, Chance, and the Future of Evolution* (Riverhead Books, New York, NY, 2017), provides an informative and entertaining account of the causes and pace of evolutionary change.

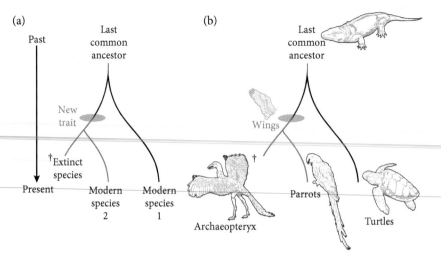

Fig. 1.3 Evolutionary trees. (a) A generic example. The oval marks the emergence or loss of a heritable trait. The dagger (†) indicates an extinct lineage. (b) A real-life example matching part a. A new trait, wings, emerged during evolution: an innovation that all birds, living and extinct, inherited.

every bifurcation represents a single species splitting into two species. The generic tree in Figure 1.3a includes two modern species and an extinct one, the latter marked by a dagger (†). The oval marks the emergence of a new trait in a species that was ancestral to both modern species 2 and the extinct species. The colored lines represent the persistence of the innovative trait in descendants of the ancestral species.

In Figure 1.3b, we provide a real-life example. A dinosaur that was the ancestor of all birds evolved wings, which an extinct bird called *Archaeopteryx* inherited. All modern birds, including parrots, also inherited wings. Therefore, we can say that wings are *homologs* (or, equivalently, *homologies*) of each other in birds. Importantly, the trait called "wings" exemplifies an evolutionary novelty. Descendants of the species marked with an oval in Figure 1.3b inherited a genetic program for developing wings, which didn't exist in species toward the trunk of the tree (upward in the diagram). The wings of other animals evolved independently: in several kinds of insects, in an extinct group of reptiles called pterosaurs (including pteranodons), and in bats. Their wings perform (or performed) the same function as a bird's, generating lift for flight, but they evolved independently from ancestors that lacked wings. So, they are not homologies; they are *analogies*. In the specialized language of biology, the term *homology* refers to ancestry and doesn't necessarily imply anything about function or overall similarity; *analogy* refers to a similarity in function without invoking a common ancestry.

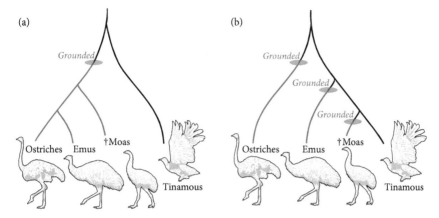

Fig. 1.4 Flightless birds. (a) A proposed evolutionary tree based on anatomical traits, which implies that flightlessness (designated as the trait grounded) evolved only once. (b) A revised tree based on DNA analysis, which shows that flightlessness evolved independently at least three times, in distinct lineages.

In fact, homologs might be quite dissimilar from each other and have very different functions;* and similarities don't necessarily indicate a homology. The reason is that similarities can develop through independent evolution, also known as *convergent evolution*, or they might be such common traits that they contribute nothing to establishing homologies. For example, a turtle's tail and a bird's brain are both composed of cells, but a tail is not homologous with a brain.

Now consider the birds in Figure 1.4. Like Figure 1.3, this illustration includes a bird capable of flight—in this case a species called the *Tinamou*. It also includes three flightless species. Flightless birds share several traits: a simpler breastbone than flying birds have, weaker chest muscles, larger hips, and stronger legs. Emus and ostriches belong to this group, as did the moa of New Zealand, which became extinct in the 13th century. In studying the evolution of flightlessness, biologists ran into a common problem: an apparent contradiction between anatomy and DNA sequences. As we illustrate in Figure 1.4a, the anatomy suggests that flightless birds descended from a common flightless ancestor because they share the traits listed above, all of which distinguish them from flying birds. Recently, however, an analysis of DNA sequences led to a different conclusion, which we depict in Figure 1.4b. Three different groups of birds lost flight independently, through convergent evolution.[17] The DNA evidence trumps similarities in anatomical traits because DNA contains a lot more information.

---

* In Chapter 5, in the section entitled "Mammals make their move," we'll present an example of a homologous structure that helps some species eat and others hear.

This example illustrates four points that can help you understand the evolution of human memory: (1) convergent evolution occurs more commonly than expected, even by experts;* (2) evolutionary changes can lead to "unnecessarily" complex and inelegant combinations of traits, such as birds with aerodynamically useless wings; (3) homologs can change their functions during evolution; and (4) when considering homologies, it is important to think in terms of extinct species in addition to modern ones. Taking these points, in turn:

- It can be difficult to determine whether a trait first evolved in an ancient common ancestor or more recently in separate lineages. For example, an ability to remember and recognize faces evolved convergently in sheep and primates, although some scientists have assumed otherwise.[18] Sheep and primates evolved from different ancestors, both of which foraged at night. In both lineages, small nocturnal animals gave rise to large diurnal descendants with big brains. And, as we'll explain in Chapter 2, their last common ancestor had a small brain, which almost certainly lacked the areas that represent faces. So, both the brain areas for face recognition and the neural representations in these areas must have evolved convergently in sheep and primates. This principle arises repeatedly in our discussion of memory. In Chapters 6–8, for example, we'll discuss several brain areas that emerged during primate evolution. We'll say that these areas and their neural representations are evolutionary innovations of primates. The existence of similar areas in other mammals doesn't contradict the idea that these areas evolved in primates. This point is important because an understanding of when and in what kind of animal these brain areas evolved can provide important clues about what they did then and what they do now.

- A common misconception about evolution is that it always produces elegant or optimal results, amenable to simple explanations. And scientists like simple explanations. Occam's razor says that scientists should always prefer the simplest idea among alternatives: the parsimony principle. But evolution doesn't produce parsimony; it produces advantages (along with neutral changes). So, the quest for simple explanations often clashes with the complexity of biological systems. Many theories of memory reduce them to two or three kinds, which has obvious attractions in terms of parsimony. But there's no good reason to favor the simplest ideas about memory. Human memory has a messy complexity that reflects a long evolutionary road traveled by ancestors very different from ourselves.

---

* You can find a discussion of convergent evolution in *Convergent Evolution: Limited Forms Most Beautiful* (MIT Press, Cambridge, MA, 2011) by George R. McGhee and in *Life's Solution: Inevitable Humans in a Lonely Universe* (Cambridge University Press, Cambridge, UK, 2003) by Simon Conway Morris.

- Homologous structures can differ in function and often do. In the chapters to come, we'll say that the function of the hippocampus changed during evolution: from storing memories that guided navigation to a role in human memories that have nothing to do with navigation. This idea seems implausible to some scientists. Yet, everyone accepts the fact that forelimbs changed their function over time: from dragging animals over dry land to propelling birds and bats through the sky. Ostriches use their wings for balance during sprints, and birds use wings for sending social signals. In fact, wings with colorful feathers first evolved in dinosaurs that were far too large and heavy to fly. So the forelimbs of dinosaurs probably changed their functions from social displays to flight, a trait inherited by the dinosaurs we call birds. If the function of forelimbs can change, why shouldn't the same principle apply to the hippocampus?

- Another common misconception is that modern animals line up in an ordered sequence, called the evolutionary scale, the scale of nature, or the *scala naturae*, with one species serving as a "precursor" to the next. When you see a drawing of several animals in a line leading to humans or read about modern monkeys having a "precursor" of language,[19] you are witnessing this misconception in action. The series of direct ancestors that we consider in this book—early vertebrates, early mammals, early primates, later primates, and humans—might seem like such as sequence, but it isn't. In each case, we refer to founding populations, not to their modern descendants.

Sloth and success

You'll need to remember what we mean by *representation* and *homology* to understand the rest of this book, but it might strike you as unlikely that a useful understanding of human memory could arise from just two concepts. To allay this fear, we quote Daniel Webster, a 19th-century American politician who, as an undergraduate, gained a reputation as both a slothful student and a highly successful one. To explain these contradictory impressions, he offered the following words of wisdom[20]:

> Many other students read more than I did and knew more than I did. But so much as I read, I made my own. When a half hour or an hour, at most, had elapsed, I closed my book and thought over what I had read. If there was anything particularly interesting or striking in the passage, I endeavored to recall it and lay it up in my memory . . . Then, if in debate or conversation afterwards, any subject came up on which I had read something, I could talk very easily so far as I had read, and then I was very careful to stop. Thus, greater credit was given me for extensive and accurate knowledge than I really possessed.

This book cannot make you an expert on human memory, but we hope that it offers some "interesting or striking" ideas to lay up in memory. If "in debate or conversation afterwards," you receive "greater credit . . . for extensive and accurate knowledge" than you deserve, what's the harm in that?

Speaking of sloth, in the next chapter we'll explain how this attribute—along with greed and cowardice—helped a small, aquatic animal survive more than 500 million years ago. From their modest brains mighty hemispheres arose, and their descendants include every vertebrate on Earth.

References

1. Milner, B., Corkin, S., & Teuber, H. L. Further analysis of hippocampal amnesic syndrome: 14-year follow-up study of H.M. *Neuropsychologia* **6**, 215–234 (1968).
2. Scoville, W. B. & Milner, B. Loss of recent memory after bilateral hippocampal lesions. *Journal of Neurology Neurosurgery and Psychiatry* **20**, 11–21 (1957).
3. Milner, B. The memory defect in bilateral hippocampal lesions. *Psychiatric Research Reports of the American Psychiatric Association* **11**, 43–58 (1959).
4. Howe, M. J. A. *Introduction to Human Memory: A Psychological Approach* (New York, NY: Harper & Row, 1970).
5. Eichenbaum, H. *The Cognitive Neuroscience of Memory: An Introduction*, 2nd edition (Oxford, UK: Oxford University Press, 2012).
6. Allen, T. A. & Fortin, N. J. The evolution of episodic memory. *Proceedings of the National Academy of Sciences U.S.A.* **110** Supplement 2, 10379–10386 (2013).
7. Murray, E. A., Wise, S. P., & Graham, K. S. *The Evolution of Memory Systems: Ancestors, Anatomy, and Adaptations* (Oxford, UK: Oxford University Press, 2017).
8. Gaffan, D. Recognition impaired and association intact in the memory of monkeys after transection of the fornix. *Journal of Comparative Physiology and Psychology* **86**, 1100–1109 (1974).
9. Mishkin, M. Memory in monkeys severely impaired by combined but not by separate removal of amygdala and hippocampus. *Nature* **273**, 297–298 (1978).
10. Meunier, M., Bachevalier, J., Mishkin, M., & Murray, E. A. Effects on visual recognition of combined and separate ablations of the entorhinal and perirhinal cortex in rhesus monkeys. *Journal of Neuroscience* **13**, 5418–5432 (1993).
11. Murray, E. A. & Mishkin, M. Object recognition and location memory in monkeys with excitotoxic lesions of the amygdala and hippocampus. *Journal of Neuroscience* **18**, 6568–6582 (1998).
12. Zola-Morgan, S., Squire, L. R., & Mishkin, M. The neuroanatomy of amnesia: amygdala-hippocampus versus temporal stem. *Science* **218**, 1337–1339 (1982).
13. Saunders, R. C., Murray, E. A., & Mishkin, M. Further evidence that amygdala and hippocampus contribute equally to recognition memory. *Neuropsychologia* **22**, 785–796 (1984).
14. Zola-Morgan, S. & Squire, L. R. Medial temporal lesions in monkeys impair memory on a variety of tasks sensitive to human amnesia. *Behavioral Neuroscience* **99**, 22–34 (1985).
15. Squire, L. R. & Zola-Morgan, S. The medial temporal lobe memory system. *Science* **253**, 1380–1386 (1991).

16. Pezzulo, G. & Cisek, P. Navigating the affordance landscape: feedback control as a process model of behavior and cognition. *Trends in Cognitive Sciences* 20, 414–424 (2016).
17. Baker, A. J., Haddrath, O., McPherson, J. D., & Cloutier, A. Genomic support for a moa-tinamou clade and adaptive morphological convergence in flightless ratites. *Molecular Biology and Evolution* 31, 1686–1696 (2014).
18. Leopold, D. A. & Rhodes, G. A comparative view of face perception. *Journal of Comparative Psychology* 124, 233–251 (2010).
19. Poremba, A. et al. Species-specific calls evoke asymmetric activity in the monkey's temporal poles. *Nature* 427, 448–451 (2004).
20. Lewis, W. *Speak for Yourself, Daniel: A Life of Webster in His Own Words* (Boston, MA: Houghton Mifflin, 1968).

2

The humble heredity
of humongous hemispheres

It's always best to start at the beginning—and all you do is follow the
yellow brick road.
                    —Glinda, the Good Witch of the North, in *The Wizard of Oz*

To understand how evolution produced human memory, you need to know something about how vertebrate brains began and changed over time. You probably
know what human brains look like, and we present our depiction in Figure 1.2.
Everything you see in Figure 1.2a is called *neocortex*. But for the first 300 million
years of vertebrate history, no such thing existed.

In contrast to the relatively recent origin of neocortex, the hippocampus is of
ancient ancestry. As you'll see, its history stretches back over 500 million years—
to the earliest vertebrates. This lengthy pedigree poses a problem that's puzzled
generations of memory scientists. In Chapter 1, we explained that as early as the
1950s scientists attributed the most sophisticated aspects of human memory to the
hippocampus, including the kinds of memory that people discuss with each other.
How could such an ancient brain area perform such a modern function? Shouldn't
new functions depend on new parts of the brain?

To answer those questions and to understand what it means for memory, it's best
to follow the Good Witch's advice and start at the beginning.

Lazy and easily frightened, yet greedy

Change is caused by lazy, greedy, frightened people looking for easier,
more profitable, and safer ways to do things. And they rarely know
what they're doing.
                    —Ian Morris, *Why the West Rules—For Now*

Not many people would be happy with a reputation for sloth, greed, and cowardice,
but the historian Ian Morris thinks that these traits shaped the most important

The Evolutionary Road to Human Memory. Elisabeth A. Murray, Steven P. Wise, Mary K. L. Baldwin,
and Kim S. Graham, Oxford University Press (2020). © Oxford University Press.
DOI: 10.1093/oso/9780198828051.001.0001

advances in human civilization. Inadvertently, his summary characterizes the life of early vertebrates remarkably well.

For animals, we tend to avoid judgmental terms like sloth, greed, and cowardice. Not always, of course: after all, sloths are called sloths for a reason; they move sluggishly and appear to lack a zest for life. Greed gets attributed, quite unfairly, to pigs; and some people hurl the epithet "pig" at anyone who takes more of something than they seem to deserve. Cowardice is a harsh word for minimizing risk. Like sloth and greed, the concept of cowardice doesn't apply to animals, with the possible exception of the Cowardly Lion.

As animals evolved, they developed various ways to choose between moving and staying put. Inaction, in the form of rest, has its place. It's not sloth, it's energy conservation. When animals rest in a place that affords safety, their behavior's not cowardice; it's a form of self-protection that promotes survival. And what might look like greed to judgmental humans is usually called motivation: the quest for things that an animal wants and likes. Evolution has produced animals that want and like what they need.

It might seem that being lazy, greedy, and easily frightened is no way to go through life, but many vertebrates do exactly that. As the ancestral population that pulled this combination together, the founding vertebrates certainly have a lot to answer for. But it would be unfair to blame them for the unsavory aspects of human nature; they were simply clinging to survival in challenging conditions.

So, what do we know about your early vertebrate ancestors and their brains? Rocks about 500–530 million years old contain fossil vertebrates,[1] but they reveal very little about the brain. Early vertebrates resembled modern fish in several ways, such as having gills and muscles on both sides of an elongated body. They could swim fast, and it's clear that these animals were highly mobile predators, feasting on slower or immobile organisms. Early vertebrates also had a head with two camera-like eyes and a brain. These traits might seem obvious: don't all animals have a head, a pair of eyes, and a brain? The answer is no: many animals lack all of these things.

The birth of the brain

In addition to studying fossils, scientists can gain insight into evolution by studying modern species. They examine the traits of each species, analyzing and contrasting as many as they can. As we explained in Chapter 1, traits are characteristics of a species, including anatomical, behavioral, and molecular ones (such as DNA sequences). When a novel trait emerges in a certain species, its descendants will tend to share that trait, but other, closely related species won't. Traits also vanish during evolution. So, by comparing similarities and differences in the traits of modern species, scientists can learn a lot about what happened during evolution.

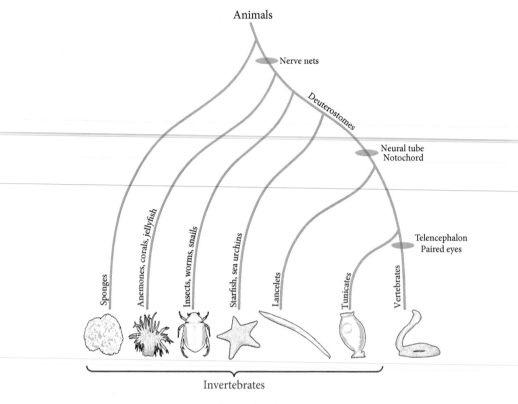

Fig. 2.1 An evolutionary tree of animals. The green lines designate deuterostomes, with blue lines for other animals. Pink ovals indicate when the traits listed to the right of each oval first emerged.

For early vertebrates, this involves identifying the heritable traits that most vertebrates share, but their closest relatives lack.

By the "closest relatives" of vertebrates, we refer to some of the animals called *deuterostomes*. As we illustrate in Figure 2.1, this group of animals—at the ends of the green lines—includes *lancelets** and *tunicates*, which share some important traits with vertebrates. A structure called the *notochord*† stiffens their body axis, at least during some phase of development, and a tube of neural tissue lies just above their notochord. Early vertebrates differed from lancelets and tunicates in several

---

* The lancelet is also known as *Amphioxus*, a creature celebrated in song and in the epigraph of this book. Regarding the epigraph, it's worth noting that we didn't exactly "come from *Amphioxus*," but instead we descended from a common ancestor that resembled *Amphioxus* in some ways but differed in others.

† The notochord, a term that comes partly from the Greek word for backside, works a little like a bony spine does in most modern vertebrates.

dramatic ways. Remarkably, they had a new kind of head, which included a pair of camera-like eyes and other sensory organs that enabled them to detect chemicals, sense their own movements, and measure gravitational forces. Their new head also supported respiration: the extraction of oxygen from water with gills. It might seem odd to consider a new kind of head as an evolutionary innovation, but the ancestors of vertebrates had nothing like it. (Other animals, such as insects, also have heads, but they evolved convergently.) In addition to a new head, early vertebrates evolved new systems for circulating blood, digesting food, moving their bodies quickly with strong muscles, and controlling the growth and metabolism of their bodies with hormones.

By contrasting modern vertebrates with their closest relatives—tunicates and lancelets—it's clear that most of the brain is a vertebrate innovation. One new part, at the front end of the brain, gave rise to human memory.

The end of the brain

The front part of the brain goes by a combination of the Greek words for end (*telos*) and brain (*enkephalos*): the *telencephalon*. All vertebrates have a brain with a prominent telencephalon, and no other kind of animal has one. So, it's reasonable to think of the telencephalon as a big breakthrough along the road to human memory.

In Figure 2.2, we highlight—in yellow, of course—our particular road through the vertebrate family tree. In *The Wizard of Oz*, the yellow brick road begins in Munchkinland and then winds its way through forests, farms, and fields to the Emerald City. Other routes, presumably with differently colored bricks, go elsewhere. In real life, divergent evolutionary roads led from a single founding vertebrate species to all of the vertebrates that live today—or ever did.

One group of vertebrates, jawless fishes, took their own road relatively early. Not only do these animals lack jaws, but they also lack the four appendages that protrude from the body of most vertebrates. In jawed fishes, paired appendages take the form of *pectoral fins* toward the front and *pelvic fins* toward the rear.*

Both types of fishes—those with and without jaws—have a telencephalon. In fact, all vertebrates do, which indicates that the telencephalon emerged before jawed and jawless vertebrates went their separate ways. So, looking back at the path evolution has taken, we can estimate when and in what kind of animal the telencephalon emerged: a jawless fish that lived before the advent of jawed fishes. This was such a significant development in the history of memory that we mark it with a signpost at the top of Figure 2.2. As these now-extinct vertebrates competed

---

* In order to minimize specialized terminology, we won't use the formal name for these animals, which is *gnathostome* from the Greek for jaw (*gnathos*) and mouth (*stoma*). Instead, we'll call them jawed fishes or jawed vertebrates. Jawless fishes are called *agnathans*.

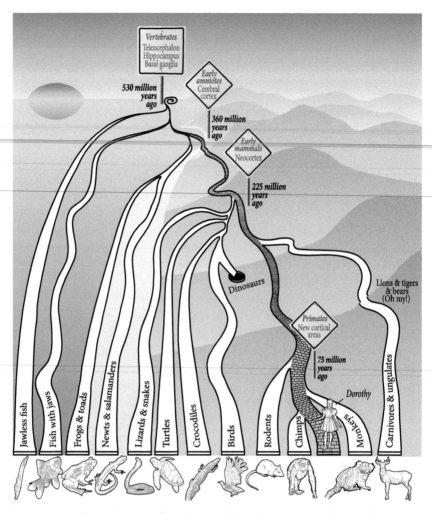

Fig. 2.2 An evolutionary tree of vertebrates. The evolutionary road traveled by modern vertebrates and one extinct group of animals: dinosaurs. Road signs mark some momentous developments along the road to human memory.

for survival more than 500 million years ago, their then-new telencephalon surely played a major role in their success, most likely by processing and storing information about the smells they encountered.[2]

The brain must end somewhere, so there is more to identifying the telencephalon than its location at the front end of the brain. Scientists have examined the brains of several vertebrate species in microscopic detail, sometimes tracing connections among brain areas and studying the expression of genes that influence brain development. This research shows that—from the start—the vertebrate brain contained a homolog of the *hippocampus* and a group of brain structures collectively called the *basal ganglia*, both of which are important for understanding human memory. We'll return to this point later.

Vertebrates gain ground

As we illustrate in Figure 2.2, the founding population of jawed vertebrates gave rise to many descendant branches. Approximately 360–420 million years ago, the pectoral and pelvic fins of a descendant species evolved into forelimbs and hindlimbs as these animals dragged themselves onto the land. Eventually, a jawed vertebrate adapted to a life entirely on land. Biologists call these animals *amniotes* because their eggs contain an *amnion*, a membrane that surrounds their embryos. Fairly early in their history, this branch of the vertebrate family tree came to another fork in the road: one branch produced dinosaurs and birds, along with modern reptiles; the other eventually gave rise to mammals.

Cortex comes alive

The vertebrate telencephalon is made up of two parts: an upper part (toward the top of the brain) and a lower part. This distinction is important for understanding memory—and merits a signpost in Figure 2.2—because in early amniotes most of the upper part evolved into the cerebral cortex: the site and source of human memory. At first, the entire cerebral cortex consisted of what's called the *allocortex*, a term we'll explain in a little while. One component of the amniote cerebral cortex is a homolog of the hippocampus; another is a homolog of the piriform cortex, also known as the olfactory (smell) cortex. In addition to brain anatomy and the pattern of gene expression during development, behavioral research supports these homologies. In Chapter 4, we'll show that removing the hippocampus causes some similar behavioral impairments in fish, mammals, and reptiles.

The "reptilian brain" fallacy

Turtles and lizards are reptiles, of course, and the idea that reptilian brains are somehow inside the human brain has a vibrant life as an internet meme. The idea of a "lizard brain" or a "reptilian brain" comes from a neurologist named Paul MacLean. His ideas remain popular, but few (if any) experts in brain evolution take them seriously. MacLean assumed that reptiles added the *basal ganglia* to a pre-existing "fish–amphibian brain" and that mammals added the hippocampus, among other cortical areas, to a pre-existing "reptilian brain." He was wrong on

(a)

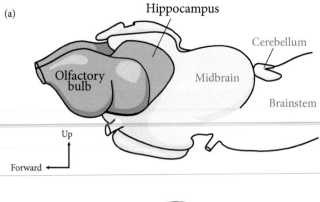

(b)

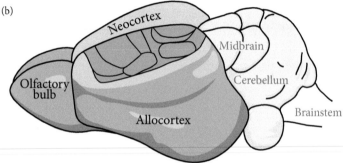

Fig. 2.3 The telencephalon. Colored parts of both drawings indicate the location of the telencephalon. (a) The brain of a lamprey, a jawless fish, as viewed from the left side, with the front of the brain to the left. The earliest vertebrates were also jawless fishes. (b) An early mammalian brain reconstructed from a comparison of modern mammalian brains. Blue-green areas mark allocortex; dark and light pink areas indicate neocortex. The lines within the pink region depict boundaries between cortical areas. The olfactory bulb processes odors and sends that information to the olfactory (smell) cortex, also known as the piriform cortex. See Figure 5.15a, in Chapter 5, for a more detailed version of part b.

both counts. As the top signpost in Figure 2.2 indicates, both the basal ganglia and the hippocampus evolved in early vertebrates, long before the appearance of either reptiles or mammals.[2] In Figure 2.3a, we illustrate the homolog of the hippocampus* in a jawless fish, the lamprey.

This point is important for understanding human memory because some scientists have argued that the basal ganglia evolved in reptiles to establish habits, a form of memory we'll discuss in Chapter 3.[†] A related idea is that the hippocampus

---

* The technical name for this part of the lamprey brain is *medial pallium*, which is homologous with the hippocampus of mammals and with the *medial cortex* of reptiles.

† In the section entitled "Instrumental memories."

evolved in mammals to establish personal memories, more commonly known as explicit or declarative memories. Because we can trace both the basal ganglia and the hippocampus much further back in evolution—to early vertebrates—we know that both ideas must be wrong. What's more, the hippocampus works in cooperation with a part of the basal ganglia (although this part usually goes by another name).* So, the idea that the basal ganglia and the hippocampus have entirely different functions—habits versus personal memories—must also be wrong.

Mammals dwell in darkness

A furnace for foraging

The third signpost in Figure 2.2, counting from the top, marks the origin of mammals about 225 million years ago. The regulation of body heat had a lot to do with their success. The ancestors of mammals needed to absorb energy from heat in the outside world, as lizards and turtles do today. Sitting on a sundrenched rock often does the trick. In contrast, mammals generate heat internally, a capacity that also evolved, convergently, in birds. Warm-blooded[†] animals have many advantages, such as the ability to forage at night and in other cool conditions, but these benefits come at a high cost. When mammals move around in the world, they burn 20–30 times more energy than a cold-blooded land animal of the same size; even at rest, they use 5–10 times more energy.[3] And this burn-rate compounds the energy costs common to all land animals, which use about ten times more energy for land locomotion than fishes use to swim.[3] So, for locomotion alone, a typical mammal uses 200–300 times more energy than your average fish. The large amount of energy that mammals can mobilize enables them to support a big brain, including large cerebral hemispheres. As we'll see in succeeding chapters, an evolutionary increase in the size of the cerebral cortex led to new kinds of memories.

With dinosaurs dominating the day, mammals scurried around in the dark and under cover for two-thirds of their time on Earth. By generating their own warmth, they could forage at night. Only after the mass extinction caused by the nasty Chicxulub, Yucatán asteroid, 66 million years ago,[4] did mammals begin to forage during the day.[5] Afterward, many mammals became larger animals with big brains.

---

* Scientists refer to this part of the basal ganglia as the septum (Latin for a divider or fence) because it seems to divide the telencephalon down the middle. The understanding that it's part of the basal ganglia comes from the work of the neuroanatomist Larry Swanson.

† The term *warm-blooded* can be a little misleading because blood can be equally warm, at times, in animals commonly called cold-blooded. Scientists instead refer to birds and mammals as *endothermic* or *homeothermic* animals. *Endothermy* refers to an internal source of heat, in contrast to *ectothermic* animals that derive their heat mainly from outside sources. *Homeothermy* refers to the ability to maintain body temperature within a set range.

The start of something big

A little earlier, we mentioned the *allocortex*. It existed from the time of early am-niotes, and we can trace its precursors to the origin of vertebrates. In contrast, the *neocortex* (meaning new cortex) arose in early mammals: another landmark event along the evolutionary road to human memory. Allocortex and neocortex differ in many ways, most notably in terms of neuronal layering. The allocortex consists of three layers, whereas the neocortex has more: six or so. In Figure 2.3b, we color allocortical areas blue-green and neocortical areas pink.

Early mammalian brains probably looked a lot like the drawing in Figure 2.3b.[6] The neocortex started small, but it eventually came to dominate the brains of most mammals. In some modern mammals, such as hedgehogs, the neocortex makes up about 15% of the brain's volume, which contrasts with about 80% in humans and whales. Early mammals probably had even less neocortex than hedgehogs do today.

Some anatomists have devised complex schemes (and names) for types of cortex that they consider to be intermediate between allocortex and neocortex. Unfortunately, these ideas have led to some serious misconceptions about cortical evolution. The neuroanatomist Fredrich Sanides invented an imaginative scheme of "evolutionary trends" and "dual origins" among cortical areas. His ideas are partly right, mostly wrong, and (unfortunately) highly influential.

Sanides thought that the cortex evolved in a sequence, with the allocortex ap-pearing first and various grades of more complexly layered areas of cortex evolving later, in a progression culminating in areas with the most complex layering. He got the first idea right; the allocortex did evolve before the neocortex. Most of the remaining cortical areas, however, emerged in an order very different than he imagined. Sanides thought that the most complexly layered parts of the cortex—sensory areas that represent skin sensations, sights, and sounds—evolved most re-cently. By comparing the brains of modern mammals, we now know that these areas were among the earliest to appear in mammals, not the latest. In Figure 2.3b, we shade the sensory areas in dark pink.

It's important to understand Sanides' mistakes because his followers, including many prominent neuroscientists, believe that the most recently evolved cortical areas are sensory areas that have relatively simple perceptual functions. The real story of cortical evolution is opposite to what Sanides imagined: the most recently evolved cortical areas have novel, complex functions that provide human memory with its distinctive properties. In Chapter 10, we'll explain why we say so.

Primates go out on a limb

In Figure 2.4, we depict the evolutionary tree of primates and our closest relatives. The most recent common ancestor of rodents, rabbits, tree shrews, and primates probably lived about 80–95 million years ago. Early primates, the lowest signpost in Figure 2.4, emerged about 74 million years ago as these animals adapted to a life confined to the fine branches of trees. Accordingly, primates and our closest relatives had been around for several million years before an asteroid strike caused the mass extinction of 66 million years ago.

Limbs and hands

The founding population of primates had two principal differences from their immediate ancestors: a reorientation of their eyes toward the front of their head; and hands and feet that could grasp objects—with fingernails and toenails instead of claws.

Both features contributed to the success of early primates. Their forward-facing eyes improved depth perception because both eyes saw more-or-less the same

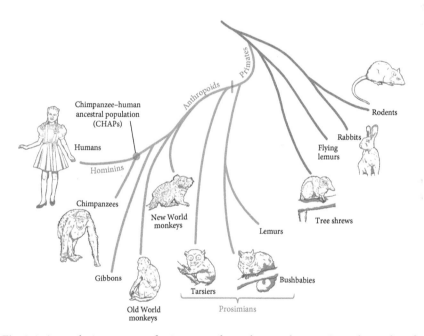

Fig. 2.4 An evolutionary tree of primates and our closest relatives. Green lines identify anthropoid primates, with blue lines for other primates and gray lines for nonprimates. The vertically oriented pink rectangle marks the last common ancestor of a group of primates called haplorhines; and the pink circle indicates the last common ancestor of humans and chimpanzees.

Adapted from Baldwin MKL, Cooke DF, Krubitzer L, Intracortical Microstimulation Maps of Motor, Somatosensory, and Posterior Parietal Cortex in Tree Shrews (*Tupaia belangeri*) Reveal Complex Movement Representations, Cerebral Cortex, 27 (2), pp. 1–18, Figure 1, doi: 10.1093/cercor/bhv329 © The Author 2016.

things. This redundancy allows the brain to translate small differences between two images, one from each eye, into distances. The ability to represent distances accurately is especially important for grasping fine branches and items on them, such as seeds, fruits, flowers, insects, and tender leaves. Leaping from place to place meant that their legs generated most of the propulsion, which freed the hands for other functions. Consequently, primates became adept at using vision to guide reaching with their arms, grasping items with their hands, and manipulating these items with precisely controlled hand and finger movements. A shorthand way of talking about primate evolution is to say that they became "visual animals." This development led to changes in their neocortex that influenced memory in profound ways. In Chapters 6–8, we'll describe new forms of memory that distinguish primates from their ancestors and from other mammals.

Arboreal life and new cortical areas

As visually guided reaching and grasping became important to primates, several new cortical areas emerged in their frontal and parietal lobes. The frontal lobe, which has the darkest shade of pink in Figure 1.2a, has two main parts: the front part has several subdivisions collectively called the *prefrontal cortex*, and most of the back part consists of areas that make up the *premotor cortex*. The parietal lobe has the lightest shade of pink in Figure 1.2a. Several new premotor, prefrontal, and parietal areas evolved in primates as part of a large-scale cortical network for selecting and guiding movements.

The cerebral cortex expanded in other mammals, too, but in primates it happened differently. As the neocortex of other mammals enlarged, the density of cortical neurons decreased so that there were fewer neurons in a given volume of cortex. Primates, however, overcame this limitation. As their cortex expanded, the density of cortical neurons remained fairly constant, presumably because of more cell division among neural stem cells. As a result of these developmental changes, primate brains have more neurons in their cortex than other mammalian brains of similar size.[7] No one knows for sure what advantages having more neurons brings, but it seems likely that this trait supports a greater degree of precision and richness in neural representations.

Premotor and parietal areas convert the observed location of an item into neural signals that control movements. They work something like a translation service. A translator at the United Nations might hear a sentence in French and repeat it in Chinese. Likewise, the primate neocortex takes the "language" of the visual world and translates it into that of the body, in which muscles generate forces that move the arm, hand, and fingers. A kernel of popcorn, for example, has a particular place in the visual world, which translates into a precise configuration of the arm, hand, and fingers that situates the thumb and index finger exactly on the kernel: it's snack time. Read Chapter 6 and you'll learn about how evolution accomplished this amazing feat of engineering.

The new prefrontal areas of early primates merit special attention. In humans, the prefrontal cortex has expanded to the point that it takes up most of the frontal lobe. However, when it began, the prefrontal cortex made up much less of the frontal lobe and had fewer areas. Todd Preuss and Pat Goldman-Rakic[8] discovered that new prefrontal areas emerged during primate evolution in at least two waves. They identified these areas by studying the brain of primates known as bushbabies (and also as galagos). As we indicate in Figure 2.4, bushbabies branched off relatively early in primate evolution. By comparing the brains of rats, bushbabies, monkeys, and humans, Preuss and Goldman-Rakic concluded that the primate frontal lobe has three groups of prefrontal areas, each of which emerged at a different time during evolution:

- One group of prefrontal areas, toward the back of the frontal lobe in primates, appeared in early mammals. Primates, rodents, rabbits, and most other mammals share these areas.
- A second group evolved in early primates, and all modern primates share these areas.
- A third group, mostly far forward in the frontal lobe, emerged later in primate evolution. Only primates on the green branches of Figure 2.4, called *anthropoids*, have these areas.

In early primates, their new prefrontal areas directed attention to items interspersed among leaves and branches, kept track of these items as the animal moved, and estimated their value based on the animal's current state. All of these adaptations required new forms of memory that humans have inherited in modified form.

Anthropoids arrive

As primate evolution went along, a group of primates split from the others about 10 million years after the first primates emerged. These animals were the founding anthropoids, a term that means "human-like." Not only are anthropoids "like" humans, some of them *are* humans. In other words, people are anthropoid primates, as are monkeys and apes.

Daylight dangers

In contrast to early primates, which foraged at night, early anthropoids foraged in daylight. This way of life has several advantages, such as the ability to see distant resources. Anthropoids inherited a visual specialization called the *fovea*, which evolved before anthropoids split from a group of primates called tarsiers.[9] In Figure 2.4, we mark the emergence of the primate fovea with a rounded pink rectangle. The fovea made vision more powerful and accurate for a small part of the world. One downside of daytime foraging is a greater risk of predation. Early anthropoids were small animals that made a tasty treat, and even after their descendants became larger animals most of them remained vulnerable to predatory birds, carnivores, and snakes. Foraging in social groups offered some protection, but the risks from predation remained high.

After anthropoids increased in size, they moved across long, horizontally oriented tree limbs, which provided highways through their foraging territory. This mode of locomotion required a lot of energy, especially when these animals needed to climb, as they often did. But by moving in this way, anthropoids could travel farther than their smaller ancestors, which tended to leap from branch to branch. Later in evolution, some anthropoids began to move mainly along the ground, which saved energy but came with a higher risk of predation.

Like modern anthropoids, the early ones probably ate a variety of foods, such as leaves, seeds, and insects. Eventually, though, many anthropoids came to rely mostly on ripe fruit for nourishment. Unfortunately for them, the availability of ripe fruit varies a great deal over time and from place to place, often with little predictability. A given kind of tree might have fruit in some of its trees but not others, and a tree that has produced fruit one year might not do so the next. For one modern anthropoid, the gray-cheeked mangabey, only about 5% of the trees in its rainforest have ripe fruit at any one time.[10] A reliance on fruit rendered anthropoids vulnerable to shortfalls in nutritional resources, and their need to forage over large territories in daylight forced them to expend energy and risk predation. During shortfalls in their preferred foods, they faced two bad choices: they could continue to seek their preferred foods, with the prospect of coming up empty; or they could settle for fallback foods. In either case, they would have to forage more often and take some serious risks every time.

The gray gets greater

The success of anthropoids depended, in part, on an expansion of the brain. Large animals tend to have large brains, so scientists usually evaluate brain size in relation to body weight. Anthropoids have larger brains, relative to their body weight, than do other primates.[11] For the most part, this expansion came from an enlargement of the neocortex, which takes up 65–80% of anthropoid brains by volume, depending on the species.[12]

Some of this expansion involved the emergence of new sensory areas, each of which processes and stores novel representations of sights, sounds, or body sensations (such as touch) in various combinations. For example, anthropoids have more visual areas than rodents, tree shrews, or bushbabies do, which indicates that several new visual areas evolved specifically in anthropoids.[6] And the visual areas that tree shrews and bushbabies do have are often less complex than their counterparts in anthropoids. For example, in some species an area called the second visual area has specialized stripes: some dedicated to color vision, some to shapes, and others to depth perception. These specializations occur in most anthropoids but not in the homologous area of tree shrews.[13] In each instance, a new kind of representation corresponds to a new form of memory, which empowers anthropoids to remember features of the world that other mammals can't. In Chapter 7, we'll discuss how some of these new areas enable anthropoids to detect distant signs of resources.

Anthropoids also have more prefrontal cortex, relative to other parts of the frontal lobe, than do other primates.[14] Earlier, we mentioned the discovery by Preuss and Goldman-Rakic that a group of new prefrontal areas emerged during anthropoid evolution. In Chapter 8, we'll explain how these new prefrontal areas help anthropoids mitigate the risk of predation, in part by reducing the frequency of bad foraging choices that expose them to danger.

Hominins develop a whale of a brain

The last common ancestor of modern apes and humans lived 18–23 million years ago, and the last common ancestor of humans and chimpanzees—the chimpanzee–human ancestral population (the CHAP)—lived 5–7 million years ago.[15] In Figure 2.4, we mark the CHAP with a pink circle.

At about this time, an era of global cooling decreased the amount of forest and increased the extent of open, grassy areas.* Many ape species became extinct, but two branches of the CHAP's descendants survived: one led to us (*hominins*); the other produced chimpanzees and bonobos. Early hominins continued to sleep and forage in trees, but they began a transition to upright walking and spent more time on the ground.

Before 2–3 million years ago, all of the CHAP's descendants had brains of roughly similar size, relative to their body weight. Afterward, the brain expanded dramatically in the hominin branch of the family tree, reaching its current size and shape somewhere between 200 000 and 600 000 years ago.† In the chimpanzee and bonobo branch, the brain expanded much less, if at all. The neocortex accounted for the lion's share of the brain's expansion in hominins, but not all of its regions

---

* Climate scientists think that a decrease in atmospheric carbon caused global cooling, which dried the subtropics 5–7 million years ago. The reason is that cool air holds less moisture, which fostered a transition from water-hungry forests to drought-resistant savannas and grasslands.
† Fossil evidence indicates that hominin brain expansion occurred in bursts, one about 2.5 million years ago and another about 1.6–1.8 million years ago.

expanded equally. Certain areas of the prefrontal, temporal, and parietal cortex expanded the most.[16-18] In Chapter 9, we'll outline some consequences of temporal-lobe and parietal-lobe expansion. Then, in Chapter 10, we'll explain how expansion of the prefrontal cortex contributed new representations that empowered an evolutionarily ancient structure, the hippocampus, to support new and uniquely human memories.

Roads reconsidered

The notion of an old brain area doing a new thing might seem incongruous, but our road metaphor helps explain this apparent paradox, at least a little. Imagine driving down a slick mountain road in foggy, treacherous conditions. You finally reach a convenience store at the foot of the mountain, get out, sell one of your tires for some cash, and buy a lottery ticket with the proceeds. Amazingly, you hit the jackpot. This windfall enables you to buy new tires, hire a skilled chauffeur, and otherwise lead a life of luxury. No one would think that Charles Goodyear invented the modern tire so that you could win the lottery someday. In this scenario, tires serve as a metaphor for the hippocampus. Originally, your tires helped you cling to a steep, slippery road and thereby survive in a challenging environment. Likewise, the hippocampus helped early vertebrates survive in their time and place. Later, your tires paved the way to additional and unrelated benefits, as the hippocampus did.*

So, here we are, on the road where Dorothy stands in Figure 2.2. One species of hominins survived, among many that lived during the past six million years. The last humans of another species, Neanderthals, died out 40 000 years ago or so. People began living in settled communities about 14 000 years ago and started to farm crops and herd domesticated animals a few thousand years after that. Permanent towns appeared 8000 years ago,[19] about the same time as people began making wine,[20] and some Sumerians started taking notes about 5000 years ago.[19] The rest, as the saying goes, is history.

In Munchkinland, the Good Witch Glinda tells Dorothy that to reach the Emerald City "all you do is follow the yellow brick road." Unfortunately, that guidance breaks down once Dorothy and Toto reach a fork in the road.

---

* Biologists refer to evolutionary developments that lay the groundwork for future changes as *exaptations*.

Befuddled by the bifurcation, they encounter a brainless bundle of straw named Scarecrow.

DOROTHY: Follow the yellow brick road? . . . Well, now which way do we go?
SCARECROW: That way is a very nice way.
DOROTHY: Who said that? . . .
SCARECROW: It's pleasant down that way, too.
DOROTHY: That's funny. Wasn't he pointing the other way?
SCARECROW: Of course, people do . . . go both ways!

But an individual can only go one way at a time, and the same goes for a population of individuals during evolution. A succession of forks in the road led to animals with a telencephalon (early vertebrates), which paved the way for animals with a cerebral cortex within the telencephalon (early amniotes), which, in turn, spawned varying configurations of neocortex within the cerebral cortex (in mammals). Along some evolutionary roads, the number of neocortical areas increased, with each area specializing in particular kinds of neural representations. In subsequent chapters, we'll examine brain specializations that we've inherited from specific ancestors, some as remote as early vertebrates and others as recent as hominin species barely different from ourselves.

Down other evolutionary roads, things turned out differently. Sometimes the number of cortical areas decreased, and some kinds of representations were probably lost. For example, the hippocampus became smaller in whales and dolphins, relative to their body weight, and in some of these species parts of the hippocampus have gone missing entirely.[21] A regression of this kind doesn't mean that whales and dolphins have an inferior brain. It's just that they don't depend as much as their ancestors did (or we do) on representations supported by the hippocampus. Presumably, whale and dolphin brains have another, different way of doing what the hippocampus does for us.

A few people can trace their family tree back a thousand years, but it's unlikely that anyone has portraits of the founding anthropoids or early primates on a mantle over the fireplace. Ever since Darwin, opponents of his theory have implied that by embracing our animal ancestry we diminish ourselves somehow. The precise opposite is the case: our evolutionary heritage enlarges our identity. The fact is that in addition to being humans we are also anthropoids, primates, mammals, amniotes, and vertebrates, among other things. Each of these identities springs from the life of an ancestral species whose survival in its time and place left a legacy in our brains and in our memories.

In addition to telling Dorothy to follow the yellow brick road, Glinda advises her to "start at the beginning." In that spirit, the next chapter deals with forms of memory that emerged in early animals, some of which exhibited behavior but didn't have a brain. The idea of behavior without brains smacks of a zombie apocalypse in which brain-dead people gnaw their way through humanity while destroying civilization. But during the evolution of life on Earth, the opposite occurred: animals with brains disrupted a world dominated by brainless organisms.

References

1. Conway Morris, S. & Caron, J.-B. A primitive fish from the Cambrian of North America. *Nature* **512**, 419–422 (2014).
2. Striedter, G. F. & Northcutt, R. G. *Brains through Time: A Natural History of Vertebrates* (Oxford, UK: Oxford University Press, 2020).
3. Ruben, J. The evolution of endothermy in mammals and birds: from physiology to fossils. *Annual Review of Physiology* **57**, 69–95 (1995).
4. Renne, P. R. et al. Time scales of critical events around the Cretaceous-Paleogene boundary. *Science* **339**, 684–687 (2013).
5. Maor, R., Dayan, T., Ferguson-Gow, H., & Jones, K. E. Temporal niche expansion in mammals from a nocturnal ancestor after dinosaur extinction. *Nature Ecology and Evolution* **1**, 1889–1895 (2017).
6. Kaas, J. H. The evolution of brains from early mammals to humans. *Wiley Interdisciplinary Review of Cognitive Science* **4**, 33–45 (2013).
7. Herculano-Houzel, S. *The Human Advantage: How Our Brain Became Remarkable* (Cambridge, MA: MIT Press, 2016).
8. Preuss, T. M. & Goldman-Rakic, P. S. Myelo- and cytoarchitecture of the granular frontal cortex and surrounding regions in the strepsirhine primate *Galago* and the anthropoid primate *Macaca*. *Journal of Comparative Neurology* **310**, 429–474 (1991).
9. Ross, C. F. in *Anthropoid Origins: New Visions* (eds. C. F. Ross & R. F. Kay) 477–537 (New York, NY: Academic/Plenum, 2004).
10. Zuberbühler, K. & Janmaat, K. in *Primate Neuroethology* (eds. M. L. Platt & A. A. Ghazanfar), pp. 64–83 (Oxford, UK: Oxford University Press, 2010).
11. Boddy, A. M. et al. Comparative analysis of encephalization in mammals reveals relaxed constraints on anthropoid primate and cetacean brain scaling. *Journal of Evolutionary Biology* **25**, 981–994 (2012).
12. Long, A., Bloch, J. I., & Silcox, M. T. Quantification of neocortical ratios in stem primates. *American Journal of Physical Anthropology* **157**, 363–373 (2015).
13. Kaas, J. H. in *The New Visual Neurosciences* (eds. J. Warner & L. Chalupa) 1233–1246 (Cambridge, MA: MIT Press, 2014).
14. Elston, G. N. et al. Specializations of the granular prefrontal cortex of primates: implications for cognitive processing. *Anatomical Record A: Discoveries in Molecular, Cellular and Evolutionary Biology* **288**, 26–35 (2006).
15. Pozzi, L. et al. Primate phylogenetic relationships and divergence dates inferred from complete mitochondrial genomes. *Molecular and Phylogenetic Evolution* **75**, 165–183 (2014).

16. Passingham, R. E. & Smaers, J. B. Is the prefrontal cortex especially enlarged in the human brain? Allometric relations and remapping factors. *Brain, Behavior and Evolution* **84**, 156–166 (2014).

17. Mars, R. B. et al. Whole brain comparative anatomy using connectivity blueprints. *eLife*, pii: e35237 (2018).

18. Hill, J. et al. Similar patterns of cortical expansion during human development and evolution. *Proceedings of the National Academy of Sciences U.S.A.* **107**, 13135–13140 (2010).

19. Scott, J. C. *Against the Grain* (New Haven, CT: Yale University Press, 2017).

20. McGovern, P. et al. Early Neolithic wine of Georgia in the South Caucasus. *Proceedings of the National Academy of Sciences U.S.A.* **114**, E10309–E10318 (2017).

21. Patzke, N. et al. In contrast to many other mammals, cetaceans have relatively small hippocampi that appear to lack adult neurogenesis. *Brain Structure and Function* **220**, 361–383 (2015).

3

Beastly brains

To have and have not

> SCARECROW: You promised . . . a real brain . . .
> WIZARD: Why, anybody can have a brain. That's a very mediocre commodity. Every pusillanimous creature that crawls on the Earth, or slinks through slimy seas has a brain!
>
> —*The Wizard of Oz*

Like everyone else, wizards make mistakes. In this case, the Great and Powerful Oz gets a simple biological fact wrong. Contrary to his opinion, many animals slink through slimy seas without a brain. For example, one species of sea anemone*— brainless as any creature could be—can swim for three minutes at a time, often in a straight line.[1] Starfish, equally brainless, slink along the sea floor at speeds up to 8 yards an hour.[2] In *The Wizard of Oz*, the Scarecrow wants a wizard to give him a brain, but animals acquired one all by themselves.

Movers and shakers—or touchy feely

Animals are aggregations—of differentiated cells. The first animals developed cells with specialized functions, along with a new genetic program for generating them. Some early animals looked something like a ball of dough. They had three layers of cells, with an opening near the bottom that took in nutrients. Eventually, these animals developed a network of interconnected neurons, called a *nerve net*, within their outermost layer. A nerve net around the opening regulated feeding behavior (ingestive movements); and, near the top of the animal, another nerve net sensed some things about the outside world and used these sensations to regulate the internal workings of the animal's body (metabolic functions).[3] A nerve net is a kind of nervous system, but it's not a brain.

Early animals split into two groups: protostomes and deuterostomes, both of which have brains. One idea about the origin of brains is that an ancestor

---

* One species of swimming anemone is called *Boloceroides mcmurrichi*. If you search for that name, you'll find YouTube® videos of its swimming prowess.

The Evolutionary Road to Human Memory. Elisabeth A. Murray, Steven P. Wise, Mary K. L. Baldwin, and Kim S. Graham, Oxford University Press (2020). © Oxford University Press.
DOI: 10.1093/oso/9780198828051.001.0001

of both protostomes and deuterostomes consolidated the two nerve nets just mentioned.[3] If so, then this ancestral species might have had a simple brain. It's clear, however, that complex brains evolved independently in protostomes and deuterostomes,[4] probably through independent consolidation of the primordial nerve nets. The difference between simple brains and complex brains is difficult to articulate because even the simplest brain has complexities of its own. But simple brains tend to have relatively few components. Take, for example, the brain of a nematode, also known as a roundworm. A satisfactory description of its brain, along with the remainder of its nervous system, takes less than 400 words.[5] Its entire nervous system has about 300 neurons,[6] compared to about 71 million neurons in the brain of a mouse,* which would take several volumes to describe adequately.

Among protostomes, the lineage leading to octopuses and squids developed a complex brain, as did the lineage that produced insects and crustaceans such as lobsters and shrimp.[4] Long after the founding deuterostomes appeared, one of their descendant species also developed a complex brain, which included a neural tube mostly corresponding to the spinal cord, and, at the front end of this tube, a brain that controlled growth and metabolism. In itself, this capacity doesn't seem so revolutionary; after all, plants control their growth and metabolism without a nervous system of any kind, let alone a centralized one with a brain. But brainy deuterostomes have one thing that plants haven't got: the ability to coordinate their active feeding movements with locomotion as they slink through slimy seas.

Moving along the seafloor has its ups and downs, literally and figuratively. Animals like corals avoid the downside by adopting a stationary life, touching and feeling the external world rather than moving through it. Earlier, we mentioned a species of sea anemone that swims, but it's the exception. Most anemone species only move occasionally and usually stay still while attached to something solid. When food floats by, these animals trap it on sticky protrusions and pull it into their bodies to extract energy. Mobile animals, in contrast, usually go to their food rather than waiting for it to come to them. When they do, they face two fundamental problems: energy expenditure and exposure to danger. An animal might bump into something sharp, wander too close to a large, hungry beast, or simply fail to obtain enough bang for its buck: like a grueling trip through snarled traffic for a stalk of celery. Despite these problems, movers and shakers dominate the animal world, not touchy-feely types like corals and anemones. Obviously, the upside of a mobile life outweighs the downside; otherwise anemones might rule the world.

---

* *The Human Advantage: How Our Brain Became Remarkable* (MIT Press, Cambridge, MA, 2016), by Suzanna Herculano-Houzel summarizes the size of brains and the number of neurons in a wide variety of species.

Fig. 3.1 Jurassic Park revisited. A gigantic anemone eats an evil lawyer in an alternative version of Michael Crichton's story about a theme park that proved to be a profound mistake in judgment.

Planet of the Anemones

What if sea anemones had adapted to life on land, as some fishes did? And what if, instead of dinosaurs, the descendants of these land-dwelling anemones had dominated the Jurassic period (about 150–200 million years ago)? It is hard to imagine that anemones would spark the human imagination as dinosaurs do, but Michael Crichton—the author of *Jurassic Park*—could have written his novel all the same. In the original story, a tyrannosaurus runs amok and eats an evil lawyer. In our version, a giant land anemone eats the lawyer instead, and in Figure 3.1 we imagine what that might look like.

Likewise, Pierre Boule could have written a novel like *Planet of the Apes* in which an astronaut lands on a world ruled by land anemones. On Rosor, in *Planet of the Anemones*, jellyfish serve as the police force, and they sting any coral or sponge who dares to challenge the authority of the reigning anemones.

These fantasies are preposterous, of course, but they serve as a reminder that life on Earth could have turned out very differently. Brains provide a powerful survival advantage, but we can imagine a world without them. Even so, animals would probably have established memories, as we explain next.

Memory in brainless animals

Although anemones don't rule the world, these animals can do a lot without a brain. In Figure 3.2, at the bottom, we use a red circle to indicate the location of sea anemones on an evolutionary tree of modern animals, one that emphasizes protostomes. Anemones, corals, and jellyfish separated from other animals relatively early in the history of animals, and sponges branched off even earlier. As we mentioned earlier, anemones have a nerve net within their body's outer layer. Sensory cells on their body surface respond to pulses of light or physical contact; and, deeper in the surface layer, other neurons connect these sensory cells to muscle cells.

We highlight sea anemones because, brainless though they are, they can establish memories of the sort that Ivan Pavlov studied in his celebrated dog. According to legend, Pavlov rang a bell and a few seconds later served his dog some food. After he'd done this several times, the dog began to drool whenever Pavlov rang his bell—before any food appeared.[*] Pavlov's dog had remembered that a ringing bell predicted food availability, and Pavlovian conditioning[†] made its debut in the history of psychology.

An experiment on anemones demonstrated that they could also establish memories of this kind, even without a brain.[7] Scientists used a flash of light instead of a ringing bell; in place of food, they substituted an electric shock. When the shock occurred alone, the anemones responded by withdrawing their tentacles and folding themselves inward to protect their most vulnerable parts. After several training trials in which a shock followed a flash of light at a constant time-lag, the anemones generated a protective movement after the flash of light rather than waiting until the shock occurred. To demonstrate that anemones establish genuine Pavlovian memories, the scientists needed to perform some additional tests. Repeated presentation of a shock can make animals highly reactive, in which case they'll respond to almost anything. Ruling out this possibility involved delivery of the shock and the light in an unpredictable order. In that circumstance, the anemones never

---

[*] As explained by Michael Specter (https://www.newyorker.com/magazine/2014/11/24/drool), Pavlov never used a bell. He used a metronome and other devices that offered more precise control over sensory stimuli.
[†] Also known as classical conditioning.

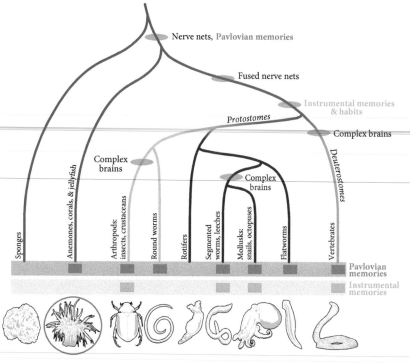

Fig. 3.2 An evolutionary tree of selected animals, with an emphasis on protostomes. Red rectangles indicate lineages known to establish Pavlovian memories. Orange rectangles do the same for instrumental memories. Pink ovals indicate possibilities for the origin of these forms of memory, along with other traits. Light blue lines indicate one major branch of the protostome family tree: molting protostomes. Dark, blue-black lines mark the other major branch: protostomes with tentacles around their mouths. We use the green line for one deuterostome lineage, vertebrates, and medium-blue lines designate other animals as well as ancestral protostomes. The red circle highlights sea anemones and their location on the animal family tree.

responded to the light. The upshot is that anemones can establish Pavlovian memories with no more brains than a scarecrow.

Anemones are naturally brainless, but headless insects can also establish new memories.[8,9] In a classic experiment, headless cockroaches or locusts had a leg placed near (but not in) saltwater. A thin wire inserted into the insect's leg completed an electrical circuit so that whenever its "foot" touched the saltwater, the animal's body received an electric shock. In the absence of a head, its leg remained under the control of a cluster of neurons called a *ganglion*, which could terminate the shock by activating muscles that pulled the "foot" away from the saltwater and kept it away. After several shocks, the ganglion formed a memory. Unlike a simple reflex, which would have jerked the "foot" from the fluid, the ganglion's memory maintained the leg in a steady posture that avoided shocks altogether.

These two brainless-memory experiments differed in an important way. The anemone's Pavlovian memory didn't enable it to avoid a shock, but the insect's memory could. The insect's kind of learning is said to result in an *instrumental memory* because an action—in this case keeping part of the body away from saltwater—was instrumental in avoiding the shock. Both types of memory generate predictions: Pavlovian memories do so on the basis of sensations; instrumental memories do so based on actions. Put another way, both types of memory record an *association*—a linkage or connection—between some initial event and a follow-on event called an *outcome*. Outcomes, in this sense, can be either beneficial or detrimental, and scientists usually use foods or fluids as beneficial (appetitive) outcomes or electric shocks as detrimental (aversive) ones. Pavlovian memories record associations between sensations and outcomes; instrumental memories record associations between actions and outcomes. In the remainder of this chapter, we'll discuss the varieties, advantages, and evolution of Pavlovian and instrumental memories.

To most people, brainless behavior is an alien concept, but in a classic movie, *To Have and Have Not*, a drunken sailor named Eddie considers the brainless behavior of bees as he sizes up Slim, a woman new to their community:

EDDIE: Say, was you ever bit by a dead bee? . . .

SLIM: Were you?

EDDIE: . . . You know, you got to be careful of dead bees . . . 'cause if you step on them they can sting you just as bad as if they was alive, especially if they was kind of mad when they got killed. I bet I been bit a hundred times that way.

Brainless bees sting; a brainless Scarecrow sings; and brainless animals remember things. It seems likely that the first animals to evolve nerve nets could establish memories, too.

Pavlovian memories

Varieties of Pavlovian memory

Pavlovian memories* come in many varieties. Sometimes, the memory that something good follows a particular sensation motivates an animal to move toward it.

---

* The word *memory* has many meanings. In this book, it refers to stored information acquired during an individual's lifetime. Through memories, past events influence neural systems and behavior. Memories, in this sense, exclude instincts and reflexes because these genetically encoded traits don't depend on an individual's experience, although experience can affect them. As we'll see in Chapter 10, the phrase *human memory* has several additional implications.

For example, if a green light indicates that food will arrive shortly, that prediction will draw a hungry animal toward the light, a behavior called *Pavlovian approach*. Pavlovian memories can also generate licking and chewing movements, the secretion of digestive fluids, and salivation, all of which prepare an animal to consume food. When Pavlovian memories predict pain, they can produce defensive responses such as freezing or a swift skedaddle. Pavlovian memories also produce arousal responses such as a gasp, a rapid heartbeat, and an increase in blood pressure.

Advantages of Pavlovian memory

The advantages of Pavlovian memories seem fairly obvious; they enable animals to prepare for beneficial and harmful events before they occur. Pavlovian predictions enable earlier and more effective behavioral responses. The sensations that trigger these responses can be neutral at first, in that they neither harm nor benefit an animal on their own.

Evolution of Pavlovian memory

We use red rectangles in Figure 3.2 to mark lineages known to establish Pavlovian memories. Protostomes in this category include flatworms, sea slugs, garden slugs, octopuses, cuttlefish, roundworms, segmented worms, and various arthropods, such as honeybees, fruit flies, and crustaceans (lobsters and shrimp, for instance). Gaps indicate a lack of testing, as far as we know, not necessarily the absence of Pavlovian learning. For simplicity, this evolutionary tree reduces vertebrates to a single representative, but all vertebrates can establish memories through Pavlovian conditioning.[10]

Given the wide distribution of species capable of establishing Pavlovian memories, it seems that virtually any network of nerve cells can do so. So, either this capacity evolved in very early animals, as indicated in Figure 3.2, or it emerged independently—but also early—in two or more lineages.

Instrumental memories

Varieties of instrumental memory

Like Pavlovian memories, instrumental memories come in several varieties. They not only permit animals to predict the general kind of outcome that should follow an action, but also specific outcomes, such as the availability of a sweet, ripe berry. It's one thing to predict that some desiccated, tasteless biscuit might be in the offing, another to anticipate strawberry ice cream or dark chocolate. Here, we use the word *outcome* to refer to any possible product of an action: good or bad.

Some instrumental memories simply record associations between actions and outcomes; others add sensations that are associated with them. So, without prompting, a dog (let's call her Daisy) might wake up thirsty from her fourth nap of the day and bark. In this case, Daisy would rely on the expectation, based on past experience, that her owner and best friend will fill up her water bowl after hearing her bark. Daisy's instrumental memory consists of an association between an action (her barking) and a predicted outcome (fluid availability): a woof–water association. Alternatively, Daisy might incorporate a sensation into an action–outcome association. She might, for example, remember that the whooshing sound made by an open faucet means that her owner is already filling her water bowl. In this case, a brisk walk to the bowl's usual location should provide a good drink: a whoosh–walk–water association. (At the same time, Daisy will have learned an association between the faucet's whooshing sound and the same predicted outcome: a Pavlovian whoosh–water association.) Not only do these different forms of memory coexist in the brain, but they also interact in various ways. For example, the Pavlovian whoosh–water association helps Daisy learn the instrumental whoosh–walk–water association, a process called Pavlovian-to-instrumental transfer. In the specialty literature, scientists describe instrumental memories in terms of stimuli, responses, and outcomes. A whoosh–walk–water association exemplifies a stimulus–response–outcome memory.

Importantly, instrumental memories can update the value of an outcome from moment to moment based on an individual's current needs and internal state. For example, eating a lot of one kind of food decreases its value. After consuming a large container of ice cream, the subjective value of more ice cream declines sharply. We'll return to this process, called *selective satiation*, later in this chapter.

In addition to memories that predict outcomes, instrumental learning produces another kind of memory. With repeated experience, instrumental memories can become so strong and so engrained that an animal will make a reflex-like response to some sensation. When it does, the predicted outcome doesn't matter any longer, and a *habit* is born. In laboratory experiments, habits are defined* as behaviors that an animal makes regardless of any predicted outcome.[11]

In Chapter 2,† we explained that some experts consider habits to be "primitive" behaviors that depend on a set of brain structures called the basal ganglia, and we pointed out why this popular idea must be wrong. One reason, among many, is that the "habit theory" of the basal ganglia overemphasizes one function of the basal

---

* The term *habit* has several conflicting meanings in psychology and biology. In this chapter, we use the meaning from a particular school of psychology: animal learning theory. In cognitive psychology, in contrast, a habit is something you can do while attending to something else. And in biology, the word habit refers to a genetically determined characteristic. Many rodents, for example, have the habit of living underground: a fossorial habit, which means that they hang around with fossils sometimes. Even trees have a habit: growing very tall.

† In the section entitled "The 'reptilian brain' fallacy."

ganglia at the expense of many others, including behaviors that—unlike habits—depend directly upon predicted outcomes.[12]

Despite this misconception, habits are important to behavior. The memories that generate habits start out weak and have little influence over behavior at first. Eventually, however, they become strong enough to generate an action before the brain can predict an outcome or its value. When this happens, behaviors become insensitive to what might occur as a result of an action: repetitive and reflex-like, but also rapid and reliable. In time, habits can dominate an animal's behavior. Consider Daisy, the dog that learned an instrumental whoosh–walk–water memory. After hundreds of repetitions, her whoosh–walk memory (a stimulus–response habit) would take control of her behavior.

Sometimes, the conflict between habits and other behaviors leads to a problem in behavioral control, especially in routine situations. One of the authors once had a particularly humiliating habit. All of the doors in his laboratory had locks controlled by a keypad that made loud clicking sounds as he entered the numbers. A four-digit number unlocked the doors, but each door required a different combination. The author's office had a combination that he had used thousands of times. So, whenever he saw a keypad of the same kind, he habitually entered the numbers that unlocked his office door. When he attempted to enter an office occupied by postdoctoral fellows, they would howl in derision as he first entered the wrong number—thus betraying the incompetent individual's identity. If you ever flipped on a light switch when fully aware that the light bulb burned out long ago or drove home directly instead of stopping for an errand as planned, you performed a habit.

Different kinds of memory often act in unison to support behavior, but, as in these examples, memories can come into conflict with each other. Typically, when this happens the strongest memory wins and controls behavior, but in Chapter 5 we'll discuss some evolutionary developments that helped early mammals manage their memories more flexibly.

Collectively, Pavlovian memories, instrumental memories, and habits can be called *conditioned reflexes*. Like reflexes in general, they occur automatically, but unlike other reflexes they are learned.

Advantages of instrumental memory

Instrumental memories enable animals to choose among several potential actions based on predicted outcomes; habits generate responses quickly and consistently. Both kinds of memory have advantages in certain circumstances. Habits save time and reduce the brain's computational burden. They provide advantages in terms of speed and reliability when the source of food or fluid rarely changes in location, amount, or likelihood. Instrumental memories work best when conditions change slowly. For example, when food has appeared in a given location 70% of the time, rather than all the time, animals adapt their foraging choices to match this percentage. They learn to choose that location 70% of the time,* which enables them to both exploit a reasonably reliable resource and explore alternatives every now and then. After all, something better might turn up. A place that only paid off 30% of the time in the past might change to 50% in the future. If so, then instrumental memories will slowly adjust the valuation of that place, and the animal will visit that location half the time.

Instrumental memories have an additional advantage over habits. As we mentioned earlier, when the memory underlying a conditioned reflex includes a predicted outcome, it can be revalued from moment to moment based on an individual's current needs and internal state. A bump or a bash might be tolerable under normal circumstances but dangerous for an injured individual. In this example, a state—injured—leads to a more negative valuation of a potentially harmful outcome. Consequently, an animal in less than tip-top shape might avoid a risk that it would otherwise take. As another example, imagine that you've just eaten as much ice cream as humanly possible. Afterward, the thought of eating any more wouldn't be very appealing. *Selective satiation* of this kind devalues a predicted outcome, which leads to different choices than you might otherwise make. Habits, being insensitive to predicted outcomes, cannot provide this kind of moment-to-moment adaptability.

In addition to these advantages, conditioned reflexes help animals cope with having "too much information," also known as information overload. In their natural habitat, animals face an onslaught of sensations, and they often need to choose from an extensive repertoire of possible actions. The variety of potential outcomes is also vast. Without some way of simplifying matters, the brain could not possibly cope with the abundance of sensations and behavioral options.

The psychologist Leon Kamin discovered one way the brain copes with information overload. His experiment involved Pavlovian memories, but the same principles apply to instrumental ones. First, he established a Pavlovian memory in one group of rats, which learned that a brief burst of noise predicted a mild shock to their feet. He then contrasted the behavior of these *trained* rats with the behavior of *naive* rats that lacked these Pavlovian memories. Both groups experienced two sensations simultaneously: the same noise plus a flash of light. A shock followed. Naive rats learned that a flash of light predicted a shock, but trained rats didn't learn anything about the light. Somehow, their Pavlovian memory—that the noise predicted a shock—prevented the trained rats from learning that a flash of light also predicted the same shock. Kamin called his result the *blocking effect* or simply *blocking*. In essence, the brains of his trained rats had simplified their job by blocking redundant learning.

---

* Experts in animal learning refer to this behavior as the *matching law*.

A second way to prevent information overload involves brain mechanisms for highly specialized functions. For example, the risk of serious harm from bringing substances into the body must be balanced against the need to consume nutrients and, in land-dwelling animals, fluids. John Garcia discovered that rats can remember a particular taste that preceded nausea or other feelings of illness. These kinds of memories, called *taste aversions*, develop after a single unfortunate experience with some food or fluid. A bad night with Scotch whiskey, for example, can put someone off this distillate for the rest of time—or so we hear. Rats learn to avoid tastes and smells associated with illness.

There is a third way of dealing with information overload: *top-down attention*. For example, contestants in many sports know that they should detect exactly how a ball rotates. Most people would never notice the precise spin-rate of a ball as it approaches, but athletes know the importance of these sensations and learn to prioritize them. Such memories influence neural processing in the visual areas of cortex and accentuate neural processing that distinguishes different kinds and degrees of spin.

All three of these brain mechanisms—top-down attention, blocking, and special-purpose mechanisms like taste aversion—mitigate information overload, but they evolved at different times and work in different ways:

- Taste aversions resemble standard Pavlovian memories by linking a sensation (a particular taste or smell) to an outcome (a subsequent feeling of illness). They differ, however, in several ways. Taste aversions require only one experience; can be established only for particular kinds of sensations (tastes and smells); and develop even when several hours intervene between a sensation and its outcome. In contrast, standard Pavlovian memories require several training trials; can be established for any sensation; and usually require a brief, fixed interval between a sensation and its outcome. Via taste aversions, evolution has shaped a special form of memory that limits the ingestion of toxins. Because taste aversions are restricted to certain kinds of sensory inputs, this powerful one-trial learning mechanism both mitigates information overload and promotes survival.

- Blocking mitigates information overload by filtering out predicted sensations (called unconditioned stimuli), and experts sometimes refer to a sensation that escapes this neural filter as a *surprise*. In cognitive psychology, surprise refers to something that causes uncertainty or disbelief about something. For conditioned reflexes, "surprise" has a different and narrower meaning. "Surprise," in this sense, unleashes a mechanism that makes a sensation better able to enter into new memories. The brain has evolved to "know" that when something "surprising" occurs, there must be something to learn. In Kamin's experiment, his *trained* rats had learned that a particular noise predicted that a shock would occur in the near future. After receiving that noise as a sensory input, the brain predicted and filtered out the shock once this sensation arrived, so nothing alerted it to learn about the flash of light that accompanied the noise. In his *naive* rats, the noise did not predict a shock—these rats hadn't learned anything yet—so the shock came as a "surprise." This unpredicted sensation stimulated the brain to establish a new memory, which linked the flash of light to the shock.

- Top-down attention mitigates information overload by accentuating the processing of some sensory signals at the expense of others, and in mammals this process is mediated mainly by the neocortex. In Chapter 2, we explained that the neocortex evolved in early mammals, and it's common to contrast neocortical functions with those of other parts of the brain, collectively called *subcortical*. The subcortex includes the cerebellum and other parts of the brainstem, among other brain structures. In contrast to "surprise," which depends on a *subcortical* mechanism that *filters out* predicted sensations,[13,14] top-down attention depends on a *neocortical* mechanism that *enhances* predicted sensations.[15] The fact that some scientists have used the same term, attention, for these starkly different brain functions has caused some avoidable confusion and obscured important differences among species.

Evolution of instrumental memory

We use orange rectangles in Figure 3.2 to mark some of the lineages known to establish instrumental memories or instrumentally conditioned habits. The gaps do not indicate an inability to form these memories; they only mean that we don't know of experiments demonstrating them in a certain group of animals. Both vertebrates and protostomes can establish instrumental memories, as can both major groups of protostomes: molting protostomes (light blue lines in Figure 3.2) and tentacle-mouth protostomes (blue-black lines). Most of this research has involved mollusks and arthropods. The former includes sea hares (also known as sea slugs or *Aplysia*), snails, octopuses, and cuttlefish; the latter includes insects (such as flies) and crabs.

In one experiment, pond snails established instrumental memories that linked an action (opening a pore in their skin to take in air) to an undesirable outcome (a tap to their body), so they refrained from opening the pore.[16] In another, fruit flies learned to fly into a particular half of a box to prevent overheating.[17] In a third example, green crabs learned to extend their claw to press one of two levers to get food, and they later learned to reverse this choice when that was the right thing to do.[18]

Among vertebrates, goldfish have learned that moving to a designated location in their tank led to a blast of oxygenated water, but only when a red light shined on the tank—not when a green light did so.[19] Another group of goldfish learned to use a shallow passageway to shuttle between two tanks and thus dodge an electric shock that followed a flash of light.[20] Toads learned to hop into one arm of a two-arm maze in order to get food.[21] And crested newts learned to snap at a circle, given the choice between a circle and a triangle.[22] Snapping at the circle yielded a worm: a highly desirable outcome for a newt. Before training, the newt didn't know to snap at the circle; afterward that was exactly what it "newt."

In addition, experiments on birds too numerous to mention have demonstrated robust instrumental memories*; and, in Chapter 4, we'll summarize similar results from experiments on two reptilian species and several mammalian ones. In all likelihood, every vertebrate species can establish instrumental memories.[10]

Given the species known to establish instrumental memories, we can estimate when this trait first evolved. Two possibilities seem most likely. As we illustrate in Figure 3.2, instrumental memories might have arisen in an ancestor common to deuterostomes and protostomes. Alternatively, these two groups of mobile animals might have evolved instrumental learning convergently. Either way, instrumental memory must have emerged relatively early in the history of animal life.

Reinforcement learning reconsidered

Ancestral advantages

Pavlovian and instrumental memories provide animals with several evolutionary advantages. Based on these memories, animals can predict beneficial or detrimental outcomes, so they can prepare in advance for what's about to happen. By registering moment-to-moment changes in the value of such outcomes, animals can make choices attuned to their current needs. These ancient forms of memory also mitigate information overload and empower animal to make choices that should yield the most beneficial or least detrimental outcome among alternatives.

As we've seen, the nerve nets and ganglia of brainless animals can establish memories. So, the ability to store information is probably a general property of neural networks, rather than a discrete capacity that derives from a special-purpose system, area, or region in the brain. It's not that a particular "box" in the brain mediates memories, with different boxes performing other neural functions. It's that all neural networks can store information, and the memories that they store will reflect the specializations and properties of each network. In this sense, the history of memory goes back to the origin of neural systems. In other words, the emergence of memory didn't rely on the emergence of a brain, let alone a brain with a neocortex, or the peculiar collection of cortical areas commonly known as the

---

* *Bird Brains: An Exploration of Avian Intelligence*, by Nathan J. Emery (Princeton University Press, NJ, 2016) presents a relatively recent summary of bird psychology.

"medial temporal lobe memory system." Animals have always had memories, and these memories have always helped them survive in a competitive world. The variety of memories in any one species will reflect the number of specialized neural systems and circuits that they have. And because memory is a fundamental property of neural systems, it should come as no surprise that the memories established by evolutionarily old neural systems have many properties in common with those that depend on more recently evolved parts of the brain.

Phony parsimony

The use of just three labels—Pavlovian memories, instrumental memories, and habits—makes memory seem simple, and some scientists believe that these conditioned reflexes account for all animal memory as well as all (or nearly all) human memory.[23,24] Parsimony has its place, but it neglects an important biological principle that we explained in Chapter 1*: evolution produces advantages (or other changes), not parsimony. Given this fact, there is no good reason to believe that there are only three kinds of memory. Modern animals have inherited the kinds of memory that emerged in early animals, but by overemphasizing them some scientists create the mistaken impression that no other forms of memory exist.

An example from the evolution of language illustrates this point. The English words for mother and brother, among many others, have "homologs" in languages as diverse as Sanskrit, Latin, Greek, German, and Armenian. These words have been passed down, in modified but recognizable form, from an ancient language spoken about 15 000 years ago. This inheritance does not mean that English consists entirely of words with such an ancient pedigree. If that were so, no one could be a muggle or a death eater. According to the *Oxford Living* Dictionaries,[25] about 185 000 words entered the English language in the 20th century alone. The moral of this story is that the persistence of old words does not preclude the introduction of new ones. Likewise, the persistence of old forms of memory, which modern species have inherited from early animals, does not preclude the existence of new forms of memory that evolved in the meantime.

Research on conditioned reflexes has yielded important advances in understanding the brain and behavior, and some experts continue to hold out hope that—in the far-distant future—the principles established by this line of research can be stretched to explain all human memory. This will never happen. To cite just a few examples, conditioned reflexes will never account for metaphorical reasoning, language, your ability to read the mental states of other people, your sense of participating in events, or your penchant for imagining events and inventing

---

* In the section entitled "Homology."

stories. To account for these cognitive capacities, we need to understand new forms of memory that arose long after the era of early animals.

Similarly, some scientists have overstated the importance of Pavlovian memories in human emotion. All animals respond to sensations that predict pain or harm through defensive actions and reactions (such as a rapid escape from dangerous situations and a surge of adrenaline). The existence of these conditioned reflexes doesn't mean that animals have the same subjective experience of fear or anxiety that humans have,[26] and the same goes for other emotions. In Chapter 10, we'll suggest that a new kind of memory, one specific to humans, led to the emotions and moods that people experience.

In *The Wizard of Oz*, Dorothy and her companions follow a path that exposes them to many perils, from harassment by the Wicked Witch of the West to a food fight with some grumpy apple trees. Dorothy, for sure, isn't in Kansas anymore. But at the beginning of her time in The Land of Oz, she has no idea of the dangers ahead. All she knows is to stick to the yellow brick road and start at the beginning. All roads have to begin somewhere; and in this chapter we've sketched the beginning of the evolutionary road to human memory. Next, we'll continue our journey down that road and see something about what made journeys possible in the first place.

References

1. Josephson, R. K. & March, S. C. The swimming performance of the sea-anemone *Boloceroides*. *Journal of Experimental Biology* 44, 493–506 (1966).
2. Montgomery, E. M. & Palmer, A. R. Effects of body size and shape on locomotion in the bat star (*Patiria miniata*). *The Biological Bulletin* 222, 222–232 (2012).
3. Arendt, D., Tosches, M. A., & Marlow, H. From nerve net to nerve ring, nerve cord and brain: evolution of the nervous system. *Nature Reviews Neuroscience* 17, 61–72 (2016).
4. Northcutt, R. G. Evolution of centralized nervous systems: two schools of evolutionary thought. *Proceedings of the National Academy of Sciences U.S.A.* 109 Suppl 1, 10626–10633 (2012).
5. Hutter, H. *C. elegans*. Simon Fraser University Department of Biological Sciences. Available at: http://www.sfu.ca/biology/faculty/hutter/hutterlab/research/Ce_nervous_system.html (2008).
6. White, J. G., Southgate, E., Thomson, J. N., & Brenner, S. The structure of the nervous system of the nematode *Caenorhabditis elegans*. *Philosophical Transactions of the Royal Society of London B: Biological Sciences* 314, 1–340 (1986).
7. Haralson, J. V., Groff, C. I., & Haralson, S. J. Classical conditioning in the sea anemone, *Cribrina xanthogrammica*. *Physiology and Behavior* 15, 455–460 (1975).
8. Horridge, G. A. Learning of leg position by the ventral nerve cord of headless insects. *Proceedings of the Royal Society of London B: Biological Sciences* 157, 33–52 (1962).
9. Horridge, G. A. Learning of leg position by headless insects. *Nature* 193, 697–698 (1962).
10. Macphail, E. M. *Brain and Intelligence in Vertebrates* (Oxford, UK: Clarendon, 1982).

11. Dickenson, A. & Balleine, B. Motivational control of goal-directed action. *Animal Learning and Behavior* 22, 1–18 (1994).
12. Yin, H. H., Ostlund, S. B., Knowlton, B. J., & Balleine, B. W. The role of the dorsomedial striatum in instrumental conditioning. *European Journal of Neuroscience* 22, 513–523 (2005).
13. Kim, J. J., Krupa, D. J., & Thompson, R. F. Inhibitory cerebello-olivary projections and blocking effect in classical conditioning. *Science* 279, 570–573 (1998).
14. Krupa, D. J. & Thompson, R. F. Inhibiting the expression of a classically conditioned behavior prevents its extinction. *Journal of Neuroscience* 23, 10577–10584 (2003).
15. Desimone, R. & Duncan, J. Neural mechanisms of selective visual attention. *Annual Review of Neuroscience* 18, 193–222 (1995).
16. Lukowiak, K., Ringseis, E., Spencer, G., Wildering, W., & Syed, N. Operant conditioning of aerial respiratory behaviour in *Lymnaea stagnalis*. *Journal of Experimental Biology* 199, 683–691 (1996).
17. Putz, G. & Heisenberg, M. Memories in drosophila heat-box learning. *Learning and Memory* 9, 349–359 (2002).
18. Abramson, C. I. & Feinman, R. D. Lever-press conditioning in the crab. *Physiology and Behavior* 48, 267–272 (1990).
19. Van Sommers, P. Oxygen-motivated behavior in the goldfish, *Carassius auratus*. *Science* 137, 678–679 (1962).
20. Scobie, S. R. & Fallon, D. Operant and Pavlovian control of a defensive shuttle response in goldfish (*Carassius auratus*). *Journal of Comparative and Physiological Psychology* 86, 858–866 (1974).
21. Schmajuk, N. A., Segura, E. T., & Reboreda, J. C. Appetitive conditioning and discriminatory learning in toads. *Behavioral and Neural Biology* 28, 392–397 (1980).
22. Hershkowitz, M. & Samuel, D. The retention of learning during metamorphosis of the crested newt (*Triturus cristatus*). *Animal Behaviour* 21, 83–85 (1973).
23. Balleine, B. W. & O'Doherty, J. P. Human and rodent homologies in action control: corticostriatal determinants of goal-directed and habitual action. *Neuropsychopharmacology* 35, 48–69 (2010).
24. Holland, P. C. Cognitive versus stimulus-response theories of learning. *Learning and Behavior* 36, 227–241 (2008).
25. Oxford Living Dictionaries. Available at: https://en.oxforddictionaries.com/explore/new-vocabulary-in-the-twentieth-century
26. LeDoux, J. E. & Pine, D. S. Using neuroscience to help understand fear and anxiety: a two-system framework. *American Journal of Psychiatry* 173, 1083–1093 (2016).

4

Vertebrate voyages

Memories of maps

HUNK: Now lookit, Dorothy, you ain't using your head about Miss Gulch. You'd think you didn't have any brains at all.
DOROTHY: I have so got brains!
HUNK: Well, why don't you use them? When you come home, don't go by Miss Gulch's place—then Toto won't get in her garden and you won't get in no trouble. See.

—The Wizard of Oz

Paths of peril

When traveling through treacherous territory, it makes sense to avoid dangerous places. Yet, for some reason, Dorothy picks the route that runs right by Miss Gulch's garden. So, Hunk has a point; although Dorothy has a first-rate brain, she isn't using it very effectively, at least not while back in Kansas. Dorothy decides to ride a risky route home, which means trouble for Toto.

Vertebrate brains evolved a way to solve Dorothy's dilemma. Like other vertebrates, Dorothy knows numerous routes home, most of which avoid Miss Gulch's garden. And she has these memories because one of her distant ancestors developed a homolog of the human hippocampus. We mentioned and illustrated the hippocampus in Chapter 1,* where we sketched the history of studies implicating this brain area in sophisticated forms of human memory.

Most of what we know about the hippocampus comes from research on rodents, monkeys, and humans, but some important clues come from considering the kind of animal in which this brain area first appeared—in the context of its life and times. In Chapter 2, we said that the hippocampus emerged in early vertebrates as they evolved into highly mobile predators, guided by smells and sights. In this chapter, we'll explain that when the hippocampus first evolved, this part of the telencephalon specialized in map-like representations of an animal's home territory, which early vertebrates used to reach places that afforded food or safe

---

* In Figure 1.2d and in the section entitled "The second road."

The Evolutionary Road to Human Memory. Elisabeth A. Murray, Steven P. Wise, Mary K. L. Baldwin, and Kim S. Graham, Oxford University Press (2020). © Oxford University Press.
DOI: 10.1093/oso/9780198828051.001.0001

haven. Map-like representations also empowered these animals to reach their goals via novel routes, to avoid trouble spots, and to surmount unexpected obstacles encountered along the way.

"Orientational" problems

Neuroscientists have developed many theories about the hippocampus, but a patient with brain damage restricted to this cortical area grasped one of its functions through introspection[1]:

> The areas of my life that I find most challenging are when I am given a series of directions, remembering my way around somewhere (familiar or unfamiliar), how I got into a building and how I can get out of it again, driving somewhere not only for the first time, but many times, remembering where I left my car and how I got into the car park in the first place, which way to turn out of a car park to get home....
>
> I would prefer not to call my experiences "memory problems," they are not. This is a total misrepresentation of the damage I have. What I experience are "orientational problems."

This patient realized that she had to navigate in an unusual way:

> I check my position at regular intervals. I literally take mental photos by stopping, turning round and taking a visual snapshot. When it is time to find my way back, I rely on my mental snapshots.

We would amend her testimony slightly. Despite this patient's preference not to call her impairment a "memory problem," that's what it is. What she means, we think, is that she doesn't have a general problem with memory; instead, she has a specific problem, one caused by the loss of the neural representations that depend on her hippocampus. Because other forms of memory remain relatively normal, she has the impression that her problem doesn't involve memory at all. We would also change her label from "orientational problem" to navigational impairment.

People and animals navigate in many ways. A memorized sequence of sensations and responses —left and right turns, for example—can suffice for highly familiar routes. Simply following a route, as Dorothy does with the yellow brick road, is another way to get somewhere. Snapshot guidance, which this patient came to rely on, can lead back to a given place when everything in a visual scene lines up just as it did in a remembered situation. Keeping track of a path through the world,* like

---

* Sometimes called *ideothetic guidance*.

Fig. 4.1 Maps. A rat plans a journey to a large chunk of cheese.

counting steps or "dead reckoning" at sea, provides another way to get somewhere. But more important than all of these methods—and a major breakthrough in the evolution of vertebrates—is navigating via map-like representations in the brain.

Maps as memory

In the 1940s, Edward Tolman developed the idea that the rat's brain synthesizes its sensory and exploratory experiences to build up and store what he called *cognitive maps*. In this context, the word *cognitive* simply refers to stored information; it doesn't imply that rats are conscious or otherwise aware of these maps.* In Tolman's view, cognitive maps enable rats to escape from danger or find food by taking novel routes through a previously explored landscape. In honor of his ideas, in Figure 4.1 we imagine a rat plotting the safest route through a cat-infested, trap-laden landscape.

---

* Information without awareness can be a foreign concept. But consider a computer or the Cloud. They store information, yet no one would say they are aware or conscious of it. Science fiction has long explored the idea of conscious computers, but they remain a fantasy—at least for now.

B. F. Skinner, an enormously influential psychologist of the mid-20th century, disagreed with Tolman. He and his followers believed that rats neither need nor use map-like representations in their brains. In their opinion, animals use a simpler approach, navigating through the world by making responses to sensations based on prior experience. According to their theory, when a rat faces a choice-point, sensations trigger the memory of a previous turn the rat had made in response to the same sensations in the past, as well as whether this turn had paid off in the end. In Chapter 3, we called these memories *conditioned reflexes*. Skinner's theories came to be called *behaviorism*, a set of doctrines that dominated psychology through much of the 20th century, especially in the United States.

These days, few scientists adhere to extreme forms of behaviorism, and the conflict between behaviorism and cognitive psychology has receded into the background for the most part. Nevertheless, tension remains between scientists who prefer one tradition over the other. Some memory scientists still prefer explanations that use the language and concepts of conditioned reflexes. Others believe that these principles will never explain the most impressive characteristics of human cognition. In Chapter 3,* we listed some of the cognitive capacities that we believe will forever defy behaviorist explanations: "metaphorical reasoning, language, your ability to read the mental states of other people, your sense of participating in events, and your penchant for imagining events and inventing stories."

Memories for moving

Google maps versus cognitive maps

That list includes some remarkable cognitive feats, but it doesn't mention navigation. Navigation is important, but it ranks pretty low on a list of your most impressive intellectual faculties. You can use global positioning satellites and smartphone apps to move around, so why would you need cognitive maps?

Unfortunately for them, early vertebrates didn't have the luxury of using navigational apps or telecommunications equipment. Instead, they relied on cognitive maps for almost everything they did: feeding, self-protection, procreation, and temperature regulation included. Given the importance of navigation to early vertebrates, it would be great to study their brains. That's impossible, of course; they lived more than 500 million years ago. But scientists can gain some insight into early-vertebrate navigation through experiments on a diverse selection of their modern descendants.

---

* In the section entitled "Phony parsimony."

Amazing mazes

Experiments on goldfish[2] and turtles[3] have revealed that the hippocampus plays a crucial role in their ability to navigate. For both species, scientists taught two groups of animals to perform a navigational task and later removed the hippocampus from the animals in one group. The goldfish learned to swim through a maze to find food, and they needed to remember the food's location to get another meal the next time through the maze. The turtles performed a different task. They needed to remember and swim to the location of a single submerged platform that had food on it—one of four such platforms in a water tank. Normal animals passed these tests, but animals without their hippocampus failed, as least for a while. (After many failures, the turtles figured out a different way to navigate, one that didn't depend on the hippocampus.)

A similar result came from a study of lizards.[4] Unlike birds and mammals, they get their warmth from the outside world. So, they like to find a warm rock and sit on it for a while to warm up. In one experiment, normal lizards remembered how to navigate to a warm rock in a box that also contained three room-temperature rocks. After the researchers removed the hippocampus from these animals, they rarely found the warm rock again.

Memories in mazes

Hundreds of experiments on rodents also support the idea that the hippocampus performs a navigational function. Since the 1940s, psychologists have known that, if possible, rats find their way to a particular place in a room that has provided them with food. They go to this goal as directly as they can, including via novel routes.

In one experiment from the 1940s,[5] rats learned to turn into one arm of a T-shaped maze to find food. (In Chapter 5, we illustrate a T-maze in Fig. 5.3c.) The shape of the maze forced the rats into one side of the room or the other. Later, the scientists placed the rats in new starting locations, which they had never experienced. In this test, the rats could take any of several straight routes through the room. Most of the rats chose a path that led them into the side of the room that had contained food in the past, even when getting there required them to turn in a different direction relative to their own bodies. Experiments like this led Tolman to conclude that rats remember spatial goals in a map-like manner and navigate by using these memories. The behaviorists' idea that rats depend on conditioned reflexes has a hard time accounting for such observations.

More recent experiments on rodents have confirmed a critical role for the hippocampus in navigational memory. In a typical experiment, rats navigated through a maze with eight arms. To pass the test, they had to remember which arms they had visited recently. In another test, called the water maze, rats found themselves in a pool of opaque water and needed to find a slightly submerged platform in order to take a break from swimming. The rats could only find the platform by remembering its location in relation to visible items outside the pool. After researchers removed or inactivated the hippocampus, rats failed all of these memory tests.[6–8]

Places, paces, and paths

A population of neurons called *place cells* reveals something important about representations in the hippocampus. As we explained in Chapter 1,* neurons distribute data to a neural network via electrical pulses called neuronal activity. For place cells, an increase in neuronal activity—more pulses in a given period of time—occurs when a rat occupies a particular location in a landscape or arena. In essence, place cells represent where a rat is at.

In the 1970s, when John O'Keefe discovered place cells, their existence seemed to vindicate Tolman's all-but-forgotten idea about cognitive maps. Although behaviorist ideas dominated neuroscience and psychology at the time, O'Keefe and Lynn Nadel[9] challenged their supremacy with two related proposals: that the activity of a network of neurons represents a map of an animal's territory and its place within it; and that the hippocampus houses this network. In addition to representing a rat's current location, the activity level of most place cells also reflects the origin of a journey or its goal.[10]

Cortical areas near the hippocampus have neurons with related properties: grid cells, boundary cells, and head-direction cells. Grid cells represent a lattice-like pattern of locations within a given area; boundary cells represent the edges of an arena; and head-direction cells represent an animal's orientation. In part, vision shapes these specialized representations, which in turn sculpt place-cell activity in the hippocampus.[11] Animals can also keep track of their movements, and this record can signal when they reach a particular location, even without visual inputs.[12] These findings show that place cells get information about an animal's location from a variety of sources, which they integrate into a coherent cognitive map.

For convenience, we'll call these maps *navigational representations*. Not only do they store information about places and journeys, but they also incorporate objects and odors encountered along the way.[11] For example, some neurons in the hippocampus represent an animal's distance and direction from some object; others represent a combination of an object and its location.

Timing also plays a crucial role in navigation. A journey through the world not only involves a series of visual scenes and smells, but also the order in which an animal encounters them and their timing. In addition to place cells, some cells in

---

* In the section entitled "Representation."

the hippocampus represent time, usually in combination with places.[13] They send out a brief signal that reflects an animal's progress through a series of places or events, for example.[14] This activity informs other brain areas about the order of landmarks along the way to some goal, as well as the time intervals between them. It's like a smartphone app that not only indicates a car's current location, but also the distance to a highway exit, the time it should take to get there, and what the exit lanes will look like when the car arrives.

These findings indicate that cognitive maps provide much more information than a map like the one in Figure 4.1. So, in a sense, Figure 4.1 is a little misleading. Rather than a traditional map, it's better to think of cognitive maps as time-rich representations of an animal's home territory, the things in it, and various journeys through it.

Steering by smell

Although the original function of these representations involved navigation, vertebrates also use them to do other things. Hundreds of experiments bear on this topic, but here we summarize three of them to illustrate a few key points. The first one shows that rats without their hippocampus have trouble remembering a sequence of smells.

In Figure 4.2, we illustrate an experiment that required rats to remember a sequence of smelly cups. Each cup contained scented sand, and some also had food buried in the sand.[15] To begin with, the rats learned two sequences of six smells, illustrated at the top of Figure 4.2 as Sequence A and Sequence B. The two sequences shared smells 3 and 4, but differed in smells 1, 2, 5, and 6. Later, the rats took a memory test, which began by digging in cups with the first two odors of a sequence. Then the rats went through smells 3 and 4, after which they faced a choice between two cups with different smells: smell 5 for Sequence A (cumin in this example) and smell 5 for Sequence B (banana). They could only get more food if they chose the smell that continued the ongoing sequence, and they could only choose one of the two cups. If they had started with Sequence A, then they need to use their memory of the two smell sequences to choose smell 5A (cumin) rather than smell 5B (banana). The white bar in Figure 4.2 shows that normal rats passed this test; they scored 30% above what they could achieve by guessing, which means they chose correctly 80% of the time. The gray bar shows that removal of the hippocampus caused rats to fail the test, and the black bar confirms that before their surgery these rats performed as well as the other group of rats.

These results show that rats need their hippocampus to remember sequences of odors. In this experiment, the cups appeared in fixed locations, but a sequence of smells corresponds to what an animal would experience during a journey through its territory. During journeys to two different places, or along two routes to the

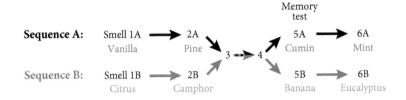

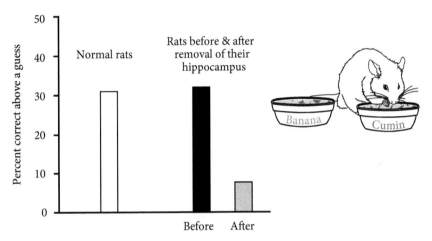

Fig. 4.2 The role of the hippocampus in sequence memory. At the top, we illustrate two six-smell sequences used in an experiment on rats. The sequences had two smells in common: smell 3 and smell 4. At the bottom, we plot the scores that two groups of rats received on a memory test for smell 5, relative to guessing. If they had only guessed and had no benefit of memory, the rats would have gotten a score of 50% correct. So, the best they could have done was 50% better than that. White bars show the average score of normal rats. Black and gray bars show average scores from a different group of rats, both before (black) and after (gray) removal of their hippocampus.

same goal, an animal will encounter a different series of smells along the way. The memory of these smells, including their order and timing, helps guide these trips.

In the 1940s, Tolman understood that rats use their cognitive maps to choose a novel route to a goal. Since his time, experiments have established that rats also use these representations for other kinds of novel choices.[16] In Figure 4.3, we illustrate such an experiment, with part a showing what rats had learned prior to a memory test. Extensive training had established two memories. First, the animals learned that after sniffing smell 1, cumin in this example, they needed to choose smell 2 (banana) rather than smell 4 (pine) to get some food. The reason was that only the cup with banana-scented sand had food buried in it. Second, the rats also learned that after sniffing smell 2 (banana) they should choose smell 3 (vanilla) rather than

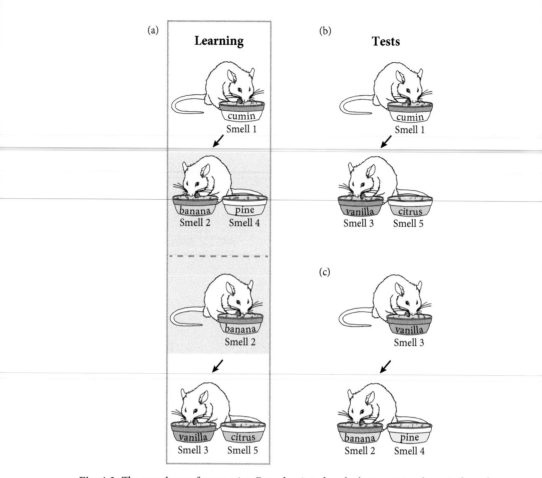

Fig. 4.3 The novel use of memories. Rats dug into bowls that contained scented sand. (a) During the learning phase of the experiment, rats learned which smell to choose after another smell. If they chose the correct cup, they could dig in it and obtain some food. If they got it wrong, they went hungry. The rats learned two consecutive associations of this kind. In this example, they should choose the banana smell after sniffing cumin; and then, in a separate learning trial (beneath the dashed line), the rats should choose the vanilla smell after sniffing banana. For illustrative purposes, the bowls have different colors and rims here (with gray rims indicating correct choices), but in the experiment they all looked exactly alike. (b) In one memory test, the rats had to combine the two memories to make a choice in a novel situation. (c) In another test, the rats had to make their choice in a novel order, based on previously learned associations among smells.

smell 5 (citrus). These memories established an order for three smells—cumin, banana, vanilla—with the middle smell linking the first one to the last. The experimenters carefully arranged these tests so that the rats couldn't solve the problem simply by remembering that a particular smell was associated with food.

In the second phase of the experiment, the rats had to use these memories to make novel choices. For instance, in their previous training they had never sniffed smell 1 (cumin) and then faced a choice between smell 3 (vanilla) and smell 5 (citrus), as we illustrate in Figure 4.3b. In essence, this test cut out the part of Figure 4.3a with the shaded background. Normal rats could put their two memories together to pass this test; after removal of their hippocampus, rats failed to do so.

The same rats also encountered smells 2 and 3 in a novel order. They had learned to sniff smell 2 (banana) and then choose smell 3 (vanilla), as we depict in the bottom half of Figure 4.3a. The rats then took the memory test that we illustrate in Figure 4.3c. After sniffing smell 3 (vanilla), they had to choose between smell 2 (banana) and smell 4 (pine). Normal rats passed this test, presumably because they knew that banana and vanilla went together and because they could use that memory in a novel situation. Removal of the hippocampus caused rats to fail this test, too.

These experiments illustrate the importance of the hippocampus for memories about sequences; and because these sequences have nothing to do with navigating through space, they also show that the hippocampus can represent relationships that aren't strictly spatial or navigational. They have such memories, in part, because the neural representations that evolved to support navigation can be useful in other ways, too.

A final example demonstrates just how far navigational memories can stray from their original functions.[17] Rats sometimes smell the breath of another rat in order to choose what to eat. That's a strange thing to say, but given two options, rats will usually choose a food with an odor resembling a fellow rat's breath. Rats seem to know that if their brethren are still breathing, then whatever they ate didn't kill them right away. In other words, bad breath beats no breath. Removal of the hippocampus disrupted these choices whenever a delay period intervened between smelling a companion's breath and choosing a food. Obviously, a social behavior of this sort couldn't have depended on breath odors when the hippocampus first emerged in early vertebrates; they lived in water and didn't breathe air. While navigating to food locations, however, some of our ancestors used their memory of a series of smells, and they presumably made choices along the way depending on what they detected. Accordingly, it was easy to adapt these representations to a social setting. Map-like representations contribute to such a broad range of functions that it is easy to lose sight of what they did for our most ancient vertebrate ancestors.

Memories in monkeys

Until the 1990s, experiments on the monkey hippocampus followed the lead of research on either the rodent or the human hippocampus. Rodent-inspired research emphasized a function in all forms of spatial memory, which we refer to here as *general spatial memory*; human-based studies focused on memories that were even more general than that, so all-encompassing that they were characterized as "global." Both ideas proved to be mistaken. In Chapter 7,* we'll summarize the research that overturned the idea that the hippocampus functions in "global" memory. In this chapter, we'll describe three experiments that refuted the idea that it functions in general spatial memory. These experiments demonstrated that the hippocampus of monkeys contributes to some forms of spatial memory but not others.

False functions

In one experiment, monkeys learned that an object in one place covered a piece of food, but an identical object somewhere else didn't. They could choose only one place on each test trial, so the monkeys quickly learned to choose the place that had food. Then the scientists switched the food to the other location. Because of the switch, this test is called a *spatial reversal*.[18,19] To pass it, the monkeys had to set aside their original memories in favor of new ones, which reflected the current situation. The first scientists to do these experiments concluded that surgical removal of the hippocampus caused monkeys to fail this test, but subsequent research showed otherwise. It turned out that inadvertent damage to nearby connections and cortical areas had caused the impairment. A better experiment, which avoided damage outside the hippocampus, showed that removal of the hippocampus, alone, had no effect on this kind of spatial memory.[20]

In another test of spatial memory, monkeys had to remember the locations of three or four objects. Later, they saw the same objects again, but only one of them appeared in its original location. That object covered food, so monkeys learned to choose it.[21,22] In an initial experiment, removal of the hippocampus seemed to cause monkeys to fail this test, but once again it had been inadvertent damage of other brain areas that had caused the impairment.[23]

A third experiment made use of the matching test that we described in Chapter 1,† modified to assess spatial memory. So, we call it the *spatial matching test*. On the first part of a trial, the monkeys saw an object in a certain place. Then, after a delay period during which they couldn't see the object, they suddenly saw

---

* In the section entitled "Bugs and blobs for brains."
† In the section entitled "The matching test."

two identical versions of it, one in its original location and the other somewhere else, both within reaching distance. The correct choice was to reach to the object's original location, which the monkeys had to remember throughout the delay period. After a few publications that led researchers down the wrong road, it again turned out that removal of the hippocampus had no effect on this kind of spatial memory.[24]

These results show that monkeys don't need their hippocampus to pass three different tests of spatial memory. Accordingly, it seems certain that the hippocampus doesn't play a *general* role in spatial memory. Instead, as we'll explain in the next section, the hippocampus supports specialized forms of spatial memory, forms that reflect its ancestral role in navigation.

Factual functions

The key experiment demonstrating this point required monkeys to navigate through a large room. From their starting point, the monkeys could see several inverted flower pots, only one of which covered food. The scientists then allowed the monkeys to explore and overturn the pots until they discovered the food, after which the monkeys returned to their starting point. Then, out of the monkeys' sight, the experimenters returned each pot to its original location, put food under the same one as previously, and waited for a variable period of time. Afterward, the monkeys could return to only one of the pots. To get another piece of food, they needed to remember where they had found food during their previous excursion and return there. Normal monkeys passed this test easily, but monkeys without a hippocampus failed badly.[25]

This experiment closely resembled the spatial matching test that we described in the previous section. In both, the monkeys needed to remember a place in order to choose it again later. The fact that the two experiments yielded contradictory results provides a powerful insight into the functions of the hippocampus. In monkeys, the hippocampus supports a specific kind of spatial memory, a kind used for navigation.

As we emphasized earlier, this conclusion doesn't mean that representations in the hippocampus contribute exclusively to navigation. The information embedded in navigational representations guides many other behaviors as well, especially those that involve sequences of events. In one study, scientists severed a neuronal pathway that connects the hippocampus, along with some nearby areas, to other parts of the brain. Afterward, these monkeys performed poorly on a task that required them to remember how recently an object had appeared in a sequence.[26] Similar findings have been reported for humans. In one experiment, people saw five images in either a fixed or random order. Activation patterns in the hippocampus carried information about both the items and their order in the sequence.[27]

Metropolis as maze

Navigation in a scanner

The most popular method for studying the human brain involves detecting significant changes in brain activation as people perform various tasks. Using this method, experimenters have asked people to solve various navigational problems in a virtual-reality cityscape. These tasks have included making judgments about the distance and direction to landmarks, the distance to a goal, any change in the distance to a goal,[28] what to do in case of a detour, the number of streets connected to a given location, the distance to the beginning of an alternative route, and the number of potential routes, among other factors.[29,30] The hippocampus and some other nearby areas became significantly activated during tasks that tax these functions, compared to control conditions.

A similar set of brain areas produced similar results for tasks related to navigation, such as appreciating the orientation of one's own head, recognizing the perspective from which someone else views a scene, identifying important landmarks in a field of view, and the timing of events.[31-34] Although it's not always obvious at first glance, these functions derive from the original role of the hippocampus, stretching back to the earliest vertebrates.

Navigation on the streets

The size of the hippocampus is linked to navigation. London taxi drivers need to learn and retain an enormous amount of finely detailed navigational information about their complex city, which is so demanding that it has its own name: the Knowledge. A study of these cabbies found that part of their hippocampus* was larger than in other professional drivers (and in other people), and the longer cabbies had been driving around London, the larger this part of the hippocampus became. The same study found that people with a larger hippocampus were also better at knowing their place in a complex array of visual landmarks.[35]

Scenes, events, and stories

Studies of patients with damage to their hippocampus have also shed light on its ancestral role in navigation. Early in this chapter,† we described one such patient,

---

* This part of the hippocampus goes by many names: dorsal, posterior, and septal hippocampus being the most common.
† In the section entitled "'Orientational' problems."

who had what she called "orientational problems." She had trouble finding the way back to her car.

Other patients, with similar brain damage, have trouble learning and remembering scenes that look like rooms they might navigate through. In one of these experiments, which we'll describe more thoroughly in Chapter 7,* people saw visual stimuli consisting of complex scenes. The experimenters presented three drawings of an identical room along with a fourth one, which depicted a different room. The task rule was to choose the "odd-one-out." Patients with damage to their hippocampus performed this task poorly for scene-like stimuli. In contrast, they easily passed otherwise identical tests involving stimuli that did not depict scenes: specifically, faces, objects, shapes, and colors.[36]

Tellingly, a similar group of patients had difficulty using their memories to imagine a novel scene or to invent stories consisting of a series of events. Their problem appeared to result from an impairment in mentally arranging their scenes and stories in a spatially coherent way.[37]

These findings show that the perception, imagination, and memory of visual scenes all depend on representations in the hippocampus and allied areas (such as the entorhinal cortex, an area that we illustrate in Figure 1.2c). These representations function in many ways that diverge from their ancestral duties, but they build on what came before. For example, the hippocampus and the entorhinal cortex represent a series of objects in much the same way that they represent a series of locations.[38] They also represent events. We just mentioned that patients with damage to their hippocampus have trouble imagining a series of events. Just as one location is distinguished from others by boundaries, such as fences between backyards, one event is distinguished from others by temporal boundaries that mark its beginning and end.[39]

Navigational representations reconsidered

In this chapter, we've reviewed research on the hippocampus (or its homolog) in goldfish, turtles, lizards, rodents, monkeys, and humans. Taken together, the results of these experiments have strengthened the idea that this brain area originally performed a navigational function. In Figure 4.4, we place a vertebrate brain near the spiral beginning of the yellow brick road and shade the homolog of the hippocampus in blue. Its ancestral role involved time-rich, map-like representations of an animal's home territory, which—from their origin in early vertebrates—came with a bonus: support for other memories, such as the sequence and timing of smells and sights.

---

* In the section entitled "Bugs and blobs for brains" and in Figure 7.2.

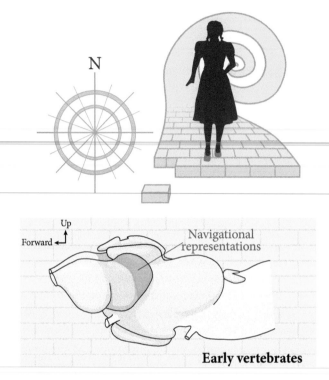

Fig. 4.4 A vertebrate innovation. The brain of a modern lamprey stands in for one of its distant vertebrate ancestors. Blue shading indicates the homolog of the hippocampus. The compass device symbolizes the ability to navigate via time-rich, map-like representations of a vertebrate's world.

- By parsing and combining these representations in various ways, an animal can build up several kinds of memories: of spatial layouts and the proximity of items to each other; of the order and recency of items in a sequence; of visual scenes that guide navigation and other behaviors; and of the multiple perspectives that come from viewing a landscape from different angles.
- The ability to remember the combination of what happened at a particular time in a particular place, commonly called an event memory,* also depends on these representations and for the same reason: navigation depends on a sequence of sensations at particular times and places.
- Much the same goes for memories of complex situations that establish a context for some behavior or perception.

---

* Most scientists use the term *episodic memory* for the representation of events. In Chapter 10, we'll discuss a more specific definition of episodic memory, which requires a sense of participation in the event.

Over time, natural selection modified the representations housed in the hippocampus, and, as that process played out, new representations arose elsewhere in the brain. In Chapter 10, we'll examine how new representations changed human memory forever. But early vertebrates didn't have an inkling of that; they needed to survive in their time and place. To understand how their navigational representations contributed to their success, imagine that you're an early vertebrate swimming for your dinner more than 500 million years ago.

In a brain that the Scarecrow would be proud to possess, your telencephalon has developed new ways to represent the world. You also have a head with exquisite odor detectors and two image-forming eyes: quite an advantage. Your deuterostome ancestors had none of these things. You have a thirst for life, but you're never thirsty because you live in water. You need food though, so despite being ensconced in a protected place, the time has come for you to go out there and get some grub. After launching yourself with a powerful swimming thrust, you shoot rapidly over a rugged escarpment and swerve swiftly and sharply to your left. At precisely this moment, you detect a distinct odor carried by a swift current, just as you have in the past. A few beats of your tail get you through the fastest part of the flow in the customary amount of time. The odor is just about the way you remember it, and a wide rock formation looms in the distance—as it always does. To the shallow-water side of those rocks, small brainless animals hang out in abundance, and they make a scrumptious meal. Today, unfortunately, you notice something unusual. Another animal—bigger than you—blocks the gap between two rocky pillars, the path you usually take. To avoid trouble, you wind your way around the pillars rather than between them. You have never gone this way before, but you can reach the shallow water all the same. It takes a little more time and effort than usual, but eventually you reach your goal. It's supper time.

At the start of this chapter, we quoted Hunk, a Kansas farmworker and Scarecrow's alter ego. His advice seems reasonable: if Dorothy can't prevent Toto from digging in Miss Gulch's garden and chasing her "nasty old cat," then taking a different route home seems like a good idea. (This is true regardless of whether Toto trespasses every day, as the wicked Miss Gulch claims, or "just once or twice a week," as Dorothy maintains.) The road past Miss Gulch's place is the fastest way, so Dorothy needs to balance the value of her time against trouble for Toto. In the next chapter, we'll explain how evolution produced a mammalian method for managing such tradeoffs: another landmark along the road to human memory.

References

1. Graham, K. S., Barense, M. D., & Lee, A. C. Going beyond LTM in the MTL: a synthesis of neuropsychological and neuroimaging findings on the role of the medial temporal lobe in memory and perception. *Neuropsychologia* 48, 831–853 (2010).

2. Rodríguez, F. et al. Spatial memory and hippocampal pallium through vertebrate evolution: insights from reptiles and teleost fish. *Brain Research Bulletin* 57, 499–503 (2002).
3. López, J. C., Vargas, J. P., Gómez, Y., & Salas, C. Spatial and non-spatial learning in turtles: the role of medial cortex. *Behavioural Brain Research* 143, 109–120 (2003).
4. Day, L. B., Crews, D., & Wilczynski, W. Effects of medial and dorsal cortex lesions on spatial memory in lizards. *Behavioural Brain Research* 118, 27–42 (2001).
5. Tolman, E. C. Cognitive maps in rats and men. *Psychological Review* 55, 189–208 (1948).
6. Olton, D. S., Collison, C., & Werz, M. A. Spatial memory and radial arm maze performance of rats. *Learning and Motivation* 8, 289–314 (1977).
7. Nadel, L. & MacDonald, L. Hippocampus: cognitive map or working memory? *Behavioral and Neural Biology* 29, 405–409 (1980).
8. Morris, R. G., Garrud, P., Rawlins, J. N., & O'Keefe, J. Place navigation impaired in rats with hippocampal lesions. *Nature* 297, 681–683 (1982).
9. O'Keefe, J. & Nadel, L. *The Hippocampus as a Cognitive Map*. (Oxford, UK: Clarendon Press, 1978).
10. Ferbinteanu, J., Shirvalkar, P., & Shapiro, M. L. Memory modulates journey-dependent coding in the rat hippocampus. *Journal of Neuroscience* 31, 9135–9146 (2011).
11. Knierim, J. J. & Hamilton, D. A. Framing spatial cognition: neural representations of proximal and distal frames of reference and their roles in navigation. *Physiological Reviews* 91, 1245–1279 (2011).
12. Chen, G., King, J. A., Burgess, N., & O'Keefe, J. How vision and movement combine in the hippocampal place code. *Proceedings of the National Academy of Sciences U.S.A.* 110, 378–383 (2013).
13. Kraus, B. J., Robinson, R. J., White, J. A., Eichenbaum, H., & Hasselmo, M. E. Hippocampal "time cells": time versus path integration. *Neuron* 78, 1090–1101 (2013).
14. Eichenbaum, H. Memory on time. *Trends in Cognitive Sciences* 17, 81–88 (2013).
15. Agster, K. L., Fortin, N. J., & Eichenbaum, H. The hippocampus and disambiguation of overlapping sequences. *Journal of Neuroscience* 22, 5760–5768 (2002).
16. Bunsey, M. & Eichenbaum, H. Conservation of hippocampal memory function in rats and humans. *Nature* 379, 255–257 (1996).
17. Bunsey, M. & Eichenbaum, H. Selective damage to the hippocampal region blocks long-term retention of a natural and nonspatial stimulus-stimulus association. *Hippocampus* 5, 546–556 (1995).
18. Jones, B. & Mishkin, M. Limbic lesions and the problem of stimulus-reinforcement associations. *Experimental Neurology* 36, 362–377 (1972).
19. Mahut, H. Spatial and object reversal learning in monkeys with partial temporal lobe ablations. *Neuropsychologia* 9, 409–424 (1971).
20. Murray, E. A., Baxter, M. G., & Gaffan, D. Monkeys with rhinal cortex damage or neurotoxic hippocampal lesions are impaired on spatial scene learning and object reversals. *Behavioral Neuroscience* 112, 1291–1303 (1998).
21. Parkinson, J. K., Murray, E. A., & Mishkin, M. A selective mnemonic role for the hippocampus in monkeys: memory for the location of objects. *Journal of Neuroscience* 8, 4159–4167 (1988).
22. Angeli, S. J., Murray, E. A., & Mishkin, M. Hippocampectomized monkeys can remember one place but not two. *Neuropsychologia* 31, 1021–1030 (1993).
23. Malkova, L. & Mishkin, M. One-trial memory for object-place associations after separate lesions of hippocampus and posterior parahippocampal region in the monkey. *Journal of Neuroscience* 23, 1956–1965 (2003).

24. Murray, E. A. & Mishkin, M. Object recognition and location memory in monkeys with excitotoxic lesions of the amygdala and hippocampus. *Journal of Neuroscience* 18, 6568–6582 (1998).

25. Hampton, R. R., Hampstead, B. M., & Murray, E. A. Selective hippocampal damage in rhesus monkeys impairs spatial memory in an open-field test. *Hippocampus* 14, 808–818 (2004).

26. Charles, D. P., Gaffan, D., & Buckley, M. J. Impaired recency judgments and intact novelty judgments after fornix transection in monkeys. *Journal of Neuroscience* 24, 2037–2044 (2004).

27. Hsieh, L. T., Gruber, M. J., Jenkins, L. J., & Ranganath, C. Hippocampal activity patterns carry information about objects in temporal context. *Neuron* 81, 1165–1178 (2014).

28. Howard, L. R. et al. The hippocampus and entorhinal cortex encode the path and Euclidean distances to goals during navigation. *Current Biology* 24, 1331–1340 (2014).

29. Spiers, H. J. & Gilbert, S. J. Solving the detour problem in navigation: a model of prefrontal and hippocampal interactions. *Frontiers in Human Neuroscience* 9, 125 (2015).

30. Javadi, A. H. et al. Hippocampal and prefrontal processing of network topology to simulate the future. *Nature Communications* 8, 14652 (2017).

31. Spiers, H. J. & Maguire, E. A. A navigational guidance system in the human brain. *Hippocampus* 17, 618–626 (2007).

32. Vass, L. K. & Epstein, R. A. Abstract representations of location and facing direction in the human brain. *Journal of Neuroscience* 33, 6133–6142 (2013).

33. Doeller, C. F., Barry, C., & Burgess, N. Evidence for grid cells in a human memory network. *Nature* 463, 657–661 (2010).

34. Auger, S. D., Mullally, S. L., & Maguire, E. A. Retrosplenial cortex codes for permanent landmarks. *Public Library of Science (PLoS) One* 7, e43620 (2012).

35. Woollett, K. & Maguire, E. A. Acquiring "the Knowledge" of London's layout drives structural brain changes. *Current Biology* 21, 2109–2114 (2011).

36. Lee, A. C. et al. Specialization in the medial temporal lobe for processing of objects and scenes. *Hippocampus* 15, 782–797 (2005).

37. Hassabis, D., Kumaran, D., & Maguire, E. A. Using imagination to understand the neural basis of episodic memory. *Journal of Neuroscience* 27, 14365–14374 (2007).

38. Garvert, M. M., Dolan, R. J., & Behrens, T. E. A map of abstract relational knowledge in the human hippocampal–entorhinal cortex. *eLife* 6, e17086 (2017).

39. Brunec, I. K., Moscovitch, M., & Barense, M. D. Boundaries shape cognitive representations of spaces and events. *Trends in Cognitive Sciences* 22, 637–650 (2018).

5

Mammalian memories

Battles in the brain

WIZARD: . . . the balloon floated down into the heart of this noble city, where I was instantly acclaimed Oz, the First Wizard de Luxe! . . . Times being what they were, I accepted the job—retaining my balloon against the advent of a quick get-away. And in that balloon, my dear Dorothy, you and I will return to the land of *E Pluribus Unum*!

—The Wizard of Oz

As everyone knows, Dorothy doesn't get home that way. When her little dog, Toto, sees a cat in the town square, he bolts from the balloon to chase it, just like he hounds Miss Gulch's "nasty old cat" back in Kansas. Dorothy then faces a dilemma: should she relinquish her ride home or lose Toto forever?

Conflicts like that arise all the time. They can be as simple as deciding whether to work out at the gym or indulge in a Big Kahuna burger; or they can be as complicated as deciding whether to leave a secure, if boring, job to join a risky startup. And, as the next section illustrates, they might even involve conflicts about controlling one's own hand.

An arm possessed by the Devil

An 84-year-old . . . woman presented to the emergency department complaining of headache and episodes of uncontrollable left hand movements. She described . . . episodes wherein her left arm moved uncontrollably as if it was groping around trying to grab herself on her body. The patient explained that while asleep she felt that "someone is trying to grab me as if someone is in bed with me." At times, she felt the need to talk to her hand or yell at it in order to command it to stop these embarrassing movements . . . [The unwanted] movements occurred while she was attempting to eat, watch television and during use of the toilet. The patient was evidently very distressed by these events and thought that her arm was "possessed by the Devil."[1]

The Evolutionary Road to Human Memory. Elisabeth A. Murray, Steven P. Wise, Mary K. L. Baldwin, and Kim S. Graham, Oxford University Press (2020). © Oxford University Press.
DOI: 10.1093/oso/9780198828051.001.0001

This patient has *alien hand syndrome*, which results from a stroke that damaged her cerebral cortex.* To grab or not to grab: you might assume that the choice is hers. Normally, her brain would resolve this conflict in accord with her intentions, but all she can do now, after her stroke, is yell at her hand and implore it to stop. The ability to resolve such battles more sensibly, without all the yelling, depends on an inheritance from early mammals.

Mammals make their move

The immediate ancestors of early mammals needed to extract heat from external sources—such as sunbaked sand, rocks, and logs—as reptiles do today. Early mammals warmed up a different way. They generated their own warmth, internally, and that meant that they could forage in the cool air of the night. As these ancestors scampered through darkness, they ceded the daytime to dinosaurs. Mammals survived and prospered, but their warm-bloodedness† came at a steep cost: 5–10 times more energy than their cold-blooded ancestors consumed at rest; 20–30 times more when on the move.[2]

Several evolutionary innovations contributed to the ability to obtain and conserve energy. For instance, only mammals have molars, which grind up seeds and grains to extract large amounts of energy. Likewise, a four-chambered heart improves the transfer of oxygen-rich (and therefore energy-rich) blood throughout the body. Hair initially served to detect contacts with the outside world, as miniature sensory organs. An enhanced ability to detect nearby objects and surfaces provided distinct advantages to early mammals as they moved around at night. Later, body hair became insulating fur that conserved heat. By 160 million years ago, an extinct mammal called *Castrocauda*—meaning beaver tail—had a body covered in a dense pelt of fur.[3] (Birds evolved warm-bloodedness convergently with mammals, accompanied by their own adaptations for extracting and conserving energy.)

As early mammals nibbled through the night, they could hear high-pitched, low-amplitude sounds better than their ancestors, a capacity that probably helped them locate noisy insects and other animals. It also opened a privileged communication channel with other members of their species. Mammalian enhancements in hearing depend, in part, on a dramatic change in the architecture of the jaw: three jaw bones morphed into middle-ear ossicles, which amplified high-pitched sounds. Big external ears helped, too.

---

* Specifically, this patient had a stroke that did most of its damage to an area called the anterior cingulate cortex, which is among the cortical areas that evolved in early mammals. We'll discuss this area again in the section entitled "Frugal or expensive?".

† In Chapter 2, we explained why scientists call these animals *endothermic* or *homeothermic* instead of warm-blooded. Other animals are called *ectothermic* or *poikilothermic*.

Early mammals also developed several new structures within their skin. Sweat glands helped them regulate body temperature; and scent glands sent social signals. In females, mammary glands provided newborns with nourishment and mammals with their name.

All of these innovations contributed to the success of mammals, but the most momentous one—at least by our way of thinking—happened inside their heads. A new kind of cortex emerged, and it changed mammalian memory forever. As we explained in Chapter 2,* the mammalian brain has two kinds of cerebral cortex: neocortex and allocortex. Because no modern reptile or bird has any neocortex, and because all mammals have a good deal of it, there can be little doubt that the neocortex made its appearance in early mammals,[3] perhaps as much as 225 million years ago. (To put this number in perspective, primates first evolved about 74 million years ago; dinosaurs became extinct about 66 million years ago; and the last common ancestor of chimpanzees and humans lived 6–7 million years ago or so.)

When the neocortex first appeared, it took up only a small portion of the cerebral cortex, which consisted mostly of allocortex. In Figure 5.1b, we illustrate what the cortex of an early mammal probably looked like.† In these extinct animals, the neocortex (pink) was small compared to the allocortex (blue-green). Although the neocortex started out small, it led to big things.

In Figure 5.1c, we use pie charts to display how much of a brain consists of neocortex. A white circle indicates an animal with no neocortex at all; the more of a circle that is pink, the more neocortex in the brain. In many mammals, including rodents, the neocortex makes up only a small part of the brain's volume: 15–30% is common. During primate evolution, about 35–45 million years ago, a primate's neocortex first reached 50% of the brain's volume,[4] eventually occupying 65–80% of it in modern monkeys, humans, and apes. Through convergent evolution, the neocortex came to dominate the brains of other mammals, too. Whales and dolphins, for example, have brains that are as much as 80% neocortex by volume; and carnivores—such as "lions and tigers and bears"—also have a lot of neocortex.

The neocortex can be divided into *core areas* (dark pink in Figs. 5.1a and b) and *ring areas* (light pink). The ring areas include parts of the neocortex called the medial prefrontal, anterior cingulate, retrosplenial, insular, orbital, and perirhinal cortex.‡ Only ring areas have a direct influence over bodily functions like

---

* In the section entitled "Vertebrates gain ground."

† Jon H. Kaas relates the story of mammalian brain evolution in his 2013 article entitled "The Evolution of Brains from Early Mammals to Humans." It can be found in the *Wiley Interdisciplinary Reviews in Cognitive Science*, volume 4, pages 33–45. As of 2018, it could be rented for 48 hours at a cost of $6 from https://onlinelibrary.wiley.com/doi/pdf/10.1002/wcs.1206.

‡ The names *medial prefrontal cortex* and *anterior cingulate cortex* reflect their locations in the brain. *Prefrontal* refers to something "in front of the front," which essentially means something like "as far forward as possible." *Medial* means toward the middle of the brain. *Cingulate* refers to a belt-like shape; *anterior* is just another word for "toward the front." The names of other prefrontal areas refer to related areas in primate brains. *Orbital* comes from the Latin word for eye socket; in this usage, the word

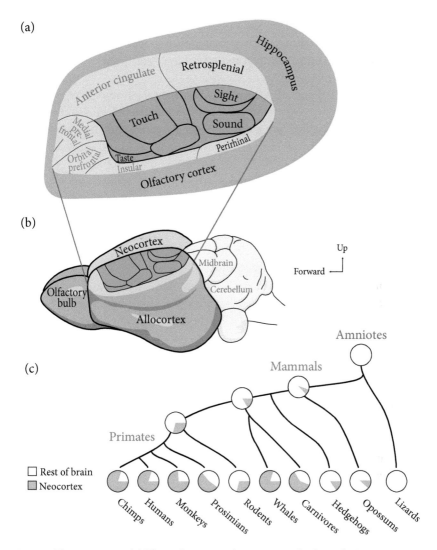

Fig. 5.1 The neocortex. (a) The early mammalian cortex, which we depict as "flattened" into two dimensions (not to scale). The neocortex is shaded pink (dark pink for its core, light pink for its ring); the allocortex is shaded blue-green. Prefrontal areas have magenta labels. (b) An early mammalian brain as viewed from the left side. (c) An evolutionary tree for selected mammals and one additional group of amniotes: lizards. The pink portion of each pie chart indicates the proportion of the brain taken up by neocortex.

orbital has nothing to do with an elliptical path around a celestial body. In primate brains, the orbital cortex seems to curve around the eye sockets from behind. The term *insular* doesn't imply narrow-mindedness, it simply refers to a part of the primate cortex that looks like an island: *insula* in Latin.

respiration and heart rate, blood pressure, and other aspects of what is called *autonomic control*. In Figure 5.1a, we depict the cortex in an idealized, "flattened" form. It's as if someone cut open a football and spread its covering flat on a table. In this distorted view, the neocortex seems larger than the allocortex. This impression is misleading for early mammals, which had a relatively small amount of neocortex, but it eventually came true for many modern mammals. The figure also allows us to illustrate the fact that most core areas specialize in representing sensations: sight, sound, and taste, along with inputs from nerves in the skin and surrounding the base of hairs (touch).*

Ring areas also represent sensory inputs. One set of *ring areas*, collectively called the insular cortex, have specializations for smells, tastes, or sensations from inside the body—in various combinations. In Chapter 1,† we mentioned another ring area, the perirhinal cortex, which specializes in the perception and memory of objects. Object identification depends mainly on vision in many species, although other kinds of sensory inputs also contribute.

It's clear that some ring areas represent sensations, but other ring areas have more enigmatic functions. The ones nearest the front of the brain, collectively called *prefrontal areas*, seem to manage battles within the brain, often based on the estimated value of alternative behaviors. In the patient with alien hand syndrome, for example, damage to one of these areas led to an unwanted winner of one such conflict, so her left hand seemed to have a life of its own.

In this chapter, when we refer to *prefrontal areas* or the *prefrontal cortex*, we mean only the small prefrontal areas that evolved in early mammals, which all mammals share by inheritance. In Chapters 6 and 8, we'll address the larger prefrontal areas that evolved specifically in primates and, in time, came to dominate the frontal lobes of apes and humans.

Inside the ring

The remainder of this chapter will take up a series of conflicts within the brain, all of which have something in common: the influence of a prefrontal area on memories that compete to control behavior. Sometimes, two prefrontal areas jockey with each other, but such battles also play out elsewhere in the brain. The prefrontal cortex manages these conflicts by generating a bias signal that favors one sort of memory at the expense of others. We begin with the area labeled as the *medial prefrontal cortex* in Figure 5.1a.

---

* Some readers might expect to see the primary motor cortex in Figure 5.1a, but this area probably evolved later: in early placental mammals rather than in early mammals.
† In the section entitled "The second road."

Expectations or efficiency?

One competition involves a battle between choices based on expectations versus habits. As we explained in Chapter 3,* *habits* are stimulus–response memories, which guide behavior quickly and without regard to expectations (called *outcomes*). *Outcome-based memories,*† in contrast, depend on associations between outcomes and something else: either sensations, or actions, or both. Animals use these memories to make choices that they expect to yield the most beneficial or least detrimental outcomes. Habits and outcome-based memories often cooperate, but they can also come into conflict.

Two parts of the original neocortex regulate these conflicts. In Figure 5.1a, we combine them into a single region labeled *medial prefrontal cortex*. To determine what its two parts do, scientists have removed or inactivated them, one at a time, in rats. They found that one part generates a bias toward habits while the other favors outcome-based memories.[5] The balance of power between these two areas determines which form of memory controls behavior.

It's important to distinguish the process of biasing competitions from the memories used to do so. If the medial prefrontal cortex rigs these competitions, what stored representations does it use to perform this function? The memories stored in the medial prefrontal cortex record whether habits or outcome-based memories have been associated with recent success (or failure), the emotional responses that follow such successes (or failures), and the sensory context prevailing at that time. Through their connections with each other and with other brain areas, the medial prefrontal areas generate a bias toward either habits or outcome-based memories. As a result, one or the other gains the upper hand in a competition to control behavior, even if it's a somewhat weaker memory.

An example, which we illustrate in Figure 5.2, helps explain this idea. Imagine a mouse emerging from its den by passing through a hole that leads to a living room. It then faces a choice between turning left toward a pantry or running right along the floor toward a table. Let's assume that the mouse has the habit of turning left to enter the pantry. Today, after finishing its pantry meal, the mouse returns to its hole, as usual. But instead of returning to its den, it chooses to explore the table, so it scampers up the drapes, jumps onto the table, and finds a tasty chunk of cheese. Tomorrow, the mouse's medial prefrontal cortex will have a representation—a memory—of the room, a route to the right and up the drapery to cheese, and the intense pleasure of eating it. This memory militates against the habitual turn toward the left (and the pantry), in favor of the right turn that might lead to cheese.

---

* In the section entitled "Instrumental memories."
† The usual term for this concept is goal-directed behavior. The idea behind this label is that the goal of an animal is to obtain something beneficial, usually food or fluid, or to avoid harm. We use the word *goal* differently, as we'll explain in Chapter 6.

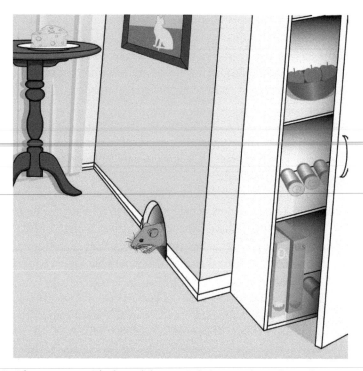

Fig. 5.2 A house mouse. The house's human inhabitants always leave the pantry door open, but they rarely leave cheese on the table.

The medial prefrontal cortex helps the mouse forage effectively by using its memories to favor one behavior or the other: choose the cheese or plunder the pantry.

The body or the world?

Another conflict involves two ways to navigate: one based on an animal's own body; the other on the outside world.* In our house-mouse scenario, we described the animal's behavior in terms of left and right turns. But the mouse can also remember the cheese's location in a world-based way, one that draws on the cognitive maps of Chapter 4. If the mouse always enters the countertop from the same place and takes the same route, a body-based approach works well; if it arrives from a different direction or has to take a novel route, a world-based approach works better, enhancing the chance of a cheesy outcome.

---

* A navigational framework based on the outside world is called allocentric (which means "other-centered"), extrinsic, or world-based. The alternatives are egocentric ("self-centered"), intrinsic, or body-based.

One way to choose between world-based and body-based navigation—and perhaps the one used by animals like lizards and frogs—is to let the strongest memories control behavior. In this case, only one representation (one memory) can guide behavior for a given stretch of time. If the dominant memory weakens sufficiently, usually due to failed foraging excursions, the alternative form of navigation can come to the fore and strengthen if it works well for a while. But the weaker of the two memories can never control an animal's foraging behavior. During mammalian evolution, the neocortex came to manage things differently. A biasing mechanism, based in the prefrontal cortex, empowered early mammals to switch rapidly between the two forms of navigation, including the form with weaker memories.

In the top half of Figure 5.3, we illustrate an experiment that studied how mammals switch between world-based and body-based navigation.[6] In their initial training, one group of rats learned to turn west at a choice point. As we depict in

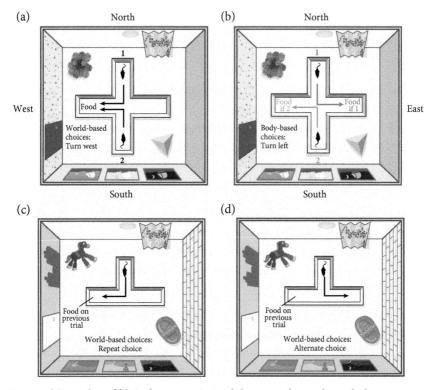

Fig. 5.3 Mazes. (a and b) A plus maze. Animals began each run through the maze at either location "1" or location "2." (a) World-based foraging: animals could pass a test by remembering to turn to the west in this example. (b) Body-based foraging: animals could pass a test by remembering to turn left in this example. (c and d) A T-maze. (c) Animals could pass a test by returning to the previous location of food. (d) Animals could pass a test by avoiding the previous location of food.

Figure 5.3a, the rats could use visible items outside the maze to guide a world-based turn to the west regardless of whether they started at point 1 or point 2. After they mastered this task, the same rats then had to learn a body-based turn instead. In Figure 5.3b, we portray a left turn at the choice point with two bent arrows: a green one for starting point 1; a blue one for starting point 2. Another group of rats learned the same two ways of navigating, but in the opposite order. Inactivation of the medial prefrontal cortex didn't prevent the rats from learning the second, conflicting form of navigation. The next day, the scientists tested the rats again. By this time, the previous day's inactivation had completely worn off. Even so, the rats used the second, newer form of navigation poorly, regardless of which one that was. They couldn't use their new memories nearly as well as rats that had an active medial prefrontal cortex throughout the experiment. The newer memory suffered because of its conflict with the older memory, a problem called *proactive interference*.

This finding shows that, at least sometimes, the medial prefrontal cortex favors recently acquired memories, giving them the boost they need to out-compete older, entrenched ones.[6,7] This bias might come from helping the brain establish or store new memories, or maybe it promotes their future use. Either way, without some help the new memories will probably lose the competition. Newer memories are typically weaker than older ones, in part because older memories interfere with establishing new ones in the first place.

As we said in the previous section, the medial prefrontal cortex uses its own representations to do what it does. In this instance, its neural networks probably record the success of a particular navigational memory and the surge of positive emotions that follow. World-based navigation depends in large part on the hippocampus, as we explained in Chapter 4. So, it's possible that the medial prefrontal cortex exerts its bias by influencing the hippocampus to dominate navigation or not, depending on circumstances.

A recent experiment revealed something important about how the medial prefrontal cortex influences the hippocampus.[8] The removal of inputs from the prefrontal cortex to the hippocampus severely weakened the neural representations of new journeys in the hippocampus. In keeping with this finding, the rats often reverted to old, obsolete routes and destinations. It's something like going to the dog food's old location in a grocery store rather than its new one.* Making a bee-line to the old place "won't feed the bulldog," as the saying goes. To avoid this mistake, the prefrontal cortex sends a signal to your hippocampus that boosts representations of a route to the dog food's new location (or suppresses alternative representations). Without this influence, your old, entrenched memory will win the competition, and you'll go to the wrong place. This explanation might also

---

* Some grocery stores reorganize their products periodically so that you'll wander around searching for what you want. That way, you'll notice other things that you might buy on impulse.

account for the ability of our house mouse to overcome its habit of turning toward the pantry in Figure 5.2. A relatively new memory—of a route to the drapes, a launching point for leaping onto the table top, and Swiss cheese—needs some help to overcome the older, stronger memory of turning left upon exiting its den. A part of the medial prefrontal cortex provides that help based on memories of its own.

Instinct or experience?

Rodents have an instinctive tendency to explore new places for food, a behavior called *spontaneous exploration* or *spontaneous alternation*. If, for example, a rat has just obtained food after running down one arm of a T-maze, it will usually switch to the other arm the next time. This innate foraging strategy is what led our house mouse to its chunk of cheese in the first place. Unfortunately, this approach can be self-defeating. When food can be found in one place and nowhere else, an animal benefits by resisting its inclination to explore elsewhere.

In an experiment exploring this behavior, rats had to return to the same arm of a maze after getting food there on the previous trial. We illustrate this task in Figure 5.3c, and its rules were similar to the *spatial matching test* that we discussed in Chapter 4.* On an initial trial, called a sample run, a rat could only enter one of the arms, let's say the west arm, which had food. On the next trial, called a test run, the rat had to choose between the east and west arms. As we illustrate in Figure 5.3c, the correct choice—for some more food—was a repeat visit to the west arm. A group of normal rats learned to pass this test without much trouble. At first, they followed their instinctive tendency and "alternated" to the east arm on the test run, but they eventually learned to return to the west arm. After scientists removed the medial prefrontal cortex from another group of rats, they failed this test.[9] After they found food in the west arm, they couldn't learn to avoid "alternating" to the east arm, which had no food.

A third group of rats faced a different problem, which we depict in Figure 5.3d. After getting food in the west arm of the maze, their task was to switch to the east arm on the next trial. They didn't need to overcome their instinctive tendency for spontaneous alternation; they only needed to make the choice that came naturally. Unlike rats that had to overcome their instincts, rats that merely had to follow them didn't need their medial prefrontal cortex to pass this test.

These findings show that the medial prefrontal cortex helps mammals favor learned behaviors over instinctive ones: another example of its ability to boost new memories to the point that they can control behavior.

---

* In the section entitled "Memories in monkeys."

Here or there?

A relatively new experimental method, called *optogenetic inactivation*, enables researchers to block direct interactions between brain structures such as the medial prefrontal cortex and the hippocampus. Unlike other methods, these interruptions can be selective for synaptic signals going in only one direction of a two-way pathway. What's more, optogenetic inactivation can silence synapses for a brief, controllable period of time.

One such experiment, on mice, used the task that we illustrate in Figure 5.3d.[10] Like the experiment just described, a sample run was followed by a test run. When the scientists temporarily blocked the direct synaptic influence of the hippocampus on the medial prefrontal cortex, mice failed this memory test, but only when the synapses were "turned off" during the sample run, when the rats first ran into one arm of the T-maze. Blocking the synapses during the test run or at other times had no effect.

Based on these results, the hippocampus seems to have sent information to the medial prefrontal cortex that established a memory about the sample run, such as the location of the food or the route to get it. After the medial prefrontal cortex received this information (during the sample run), it didn't need inputs from the hippocampus anymore. Its own representations could then generate a behavioral bias toward the appropriate goal.

That experiment revealed something about what the medial prefrontal cortex *represents*—information about the sample run—another illustrated what the medial prefrontal cortex *does* with its representations. Scientists established two different behavioral contexts* by lining boxes with wallpaper.[11] Wallpaper with red horizontal stripes created one context; blue vertical stripes another. Both boxes contained two cups with scented sand: one smelling of grapefruit; the other of geranium. In the red-striped box, only the grapefruit-scented sand concealed food; in the blue-striped box, it was the other way around. When a rat went into the box with red-striped wallpaper, some neurons in its hippocampus had especially high levels of activity as it smelled grapefruit (the food-associated scent) in a particular place within the box, with little or no activity for geranium in that place or grapefruit anywhere else. In other words, these neurons represented a specific combination of an odor, a place, and a context: the smell of grapefruit (an odor) in the back-left corner (a place) of the red-striped box (a context). When scientists inactivated the medial prefrontal cortex, neurons in the hippocampus no longer had different levels of activity for geranium and grapefruit odors in the back-left corner of the red-striped box, and so these odors no longer helped the rat find food. These results show that the medial prefrontal cortex sends a bias signal to

---

* Many things can serve as behavioral contexts. Examples include: being at home or in someone else's house; the background scene behind an object; or the brightness of a room.

the hippocampus, one that favors representations of a particular smell–place com-bination in a given context. In this case, the prefrontal cortex boosted the memory of grapefruit odor in the back-left corner of the red-striped box, and without this boost the rats couldn't deal effectively with their conflicting memories. Earlier, we said that the medial prefrontal cortex boosts the memories of new routes at the expense of older, obsolete ones.[8] The wallpaper experiment shows that the medial prefrontal cortex can also generate a bias between simultaneously valid mem-ories: in this case, of smell–place combinations that predict food in one context but not in another.

This experiment might seem artificial but consider the house mouse that we mentioned earlier. The pantry provides one foraging context, the countertop an-other. To get enough nourishment, the mouse needs to know the best places to forage in both contexts, and smells help guide the way. It's like the knowledge you might have about two different markets. Let's say that in one market, you know that the cantaloupes smell and taste great, but the tangerines stink and taste lousy; in the market across the street, it's the other way around. If you couldn't remember contexts and their meaning, you might come home with both crummy cantaloupes and terrible tangerines.

The experiments discussed so far in this chapter come from research on ro-dents, but a study of people points to something similar. Both the medial prefrontal cortex and the hippocampus generate a kind of brain wave called a *theta rhythm*, and these brain waves get in synch just as people put two established memories together to draw an inference that combines them.[12] Perhaps when two areas syn-chronize their activity, they can communicate with each other better. Regardless, it seems likely that the medial prefrontal cortex and the hippocampus interact mean-ingfully in humans, just as they do in rodents.

Frugal or expensive?

So far, we've discussed conflicts managed by the medial prefrontal cortex. Another ring area of the neocortex is called the *anterior cingulate cortex*. Research on ro-dents indicates that it mediates conflicts between actions associated with high de-grees of effort and large pay-offs, as opposed to those associated with low degrees of effort and small pay-offs. Given their high-energy life, early mammals benefited from an ability to marshal large amounts of energy when that effort promised a benefit greater than the cost.

In an experiment that explored this conflict, rats learned that by climbing over a high wall they could get to a large amount of food. They also learned that they could get a small amount of food simply by running somewhere else at ground level. Normal rats made the strenuous effort needed to surmount the barrier and get the bigger bounty. After removal of their anterior cingulate cortex, rats took the easy way out and only obtained the smaller amount of food. They acted as though the larger amount of food wasn't worth the physical effort—the sheer exertion—of climbing over the barrier.[13, 14]

These observations show that the anterior cingulate cortex exerts a bias toward expensive foraging (in terms of energy expenditure) when it conflicts with frugal foraging. According to current thinking, it does so by storing and using memories of the stupendous outcomes that should follow an extraordinary exertion. Some research on humans has examined the homologous area in the context of effortful thinking.[15] The anterior cingulate cortex became significantly more activated when people had to think long and hard about something, as opposed to engaging in a more relaxed mental activity. So, the next time you feel that something's just too much to think about anymore, at least you'll know the brain area that's letting you down: the anterior cingulate cortex. This area seems to manage conflicts of many kinds, especially when two or more competing behaviors require different amounts of mental or physical effort.

Now or not?

Two additional prefrontal areas, the orbital[14,16] cortex and the insular[17] cortex,* help resolve the conflict between urgent and patient foraging. Patient foraging enables mammals to forgo low-quality meals in favor of a bigger or better bounty later; urgent foraging yields something decent right away. To perform their functions, these ring areas store memories about the sensations that are linked to specific tastes and smells, which guide mammals toward choices that will yield the foods and fluids they want most, given the urgency of their current situation.

Patience is said to be a virtue, but by considering the dilemma that our house mouse faces, you can see that it's not as simple as that. It's not easy to decide between patience and urgency. Sacrificing a sure thing to get a something better down the road sometimes pays off but often doesn't, especially if there're other hungry mice around. Responding immediately to every opportunity can also lead to meager returns. A mouse that always does one thing or the other might not survive. The same quandary applies to human behavior. We'd all like to have infinite patience; the problem is that we want it immediately. Sometimes, though, patience is just another word for procrastination, which merely delays an attempt to achieve a goal: hardly a virtue. At the same time, an immediate response to every attention-grabbing sensation causes problems of its own. Hard as it might be to believe, some people spend too much time texting, watching cat videos on YouTube®, or

---

* Primates have additional components of their insular cortex. Rodents and primates share the area mentioned here, which takes up the front part of the primate insula.

transfixed by their Twitter® feed. Equally astonishing, these activities occasionally reduce the amount of productive work that gets done.

The orbital cortex also helps rats make foraging choices based on their current state. After gorging on their favorite food, for example, normal rats will do whatever it takes to get a different kind of food to eat; but after the surgical removal of their orbital cortex they'll stick with whatever provides their favorite food, despite having a belly full of it.[18] The orbital cortex somehow gains access to an updated memory of a food's value, a topic that we'll take up again in Chapter 6.*

Mammalian memory reconsidered

Our house mouse inherited a mighty brain from its early mammalian ancestors. More than a quarter of its brain consists of neocortex, divided into a few dozen areas, each one specialized for a particular kind of neural representation.

A high-energy life made such a large neocortex possible, and many of their new cortical areas provided early mammals with enhancements in their sensory prowess, including feeling contacts with the outside world and processing sensations from within the body, exquisite hearing, and seeing things in a way that their ancestors couldn't.

The neocortex also endowed early mammals with an ability to learn, retain, and manage contradictory memories, each appropriate in certain contexts. Obsolete memories could be retained and used in case the "good old days" returned, and new memories received a boost when it mattered most. With new representations in prefrontal parts of their ring neocortex, early mammals could strike a balance among alternative ways of surviving in a competitive world: spending versus conserving energy; making choices based on one context versus another; using new memories versus old ones (or instincts) to guide behavior; acting habitually versus seeking the best possible outcome; acting with urgency versus patience; and making choices based on the current value of a food item versus a long-term preference, among other battles in the brain. In Figure 5.4, we label their collective specialization as *bias representations*.

At the beginning of this chapter, we mentioned Dorothy's dilemma when Toto decides to hound a Siamese cat at the precise moment the Wizard's balloon lifts off from the Land of Oz. Dorothy has only a moment to decide whether to chase Toto or take a once-in-a-lifetime, one-way ride back to Kansas. Toto faces a conflict, too: the joy of a cat chase versus devotion to Dorothy. In another context, Toto's behavior wouldn't be a problem; after all, there's nothing wrong with a good cat chase

---

* In the section entitled "Deciding on desirability."

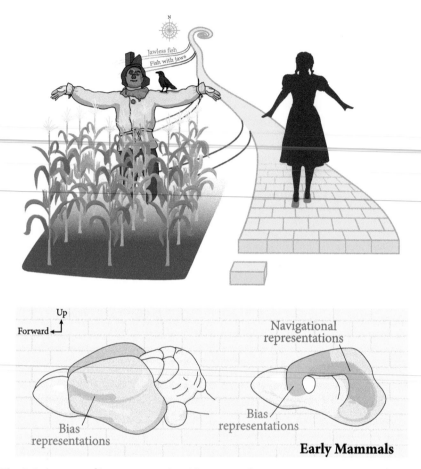

Fig. 5.4 A mammalian augmentation. Navigational representations correspond to the cognitive maps that evolved in early vertebrates, as we discussed in Chapter 4. Bias representations are the memories we've explored in this chapter. A side view of the brain is at the lower left; a view of half the brain from the middle (with the other half removed) is at the lower right. We illustrate the Scarecrow in his cornfield because he's the first companion that Dorothy meets along the yellow brick road and because he generates a bias: toward people getting the corn rather than crows. Afterward, the Scarecrow joins Dorothy on her journey to the Emerald City, just as new representations augmented older ones along the evolutionary road to human memory.

on a dull day. Toto simply picked the wrong time to indulge himself. In the event, both Toto and Dorothy resolve their inner conflict by drawing on a legacy from early mammals: Toto chases the cat; Dorothy chases Toto; and the Wizard returns to Kansas alone. Happily, Dorothy recognizes her self-sufficiency; she can use her ruby slippers to get home all on her own—with Toto in tow.

Conflicts among contradictory memories resemble the choice a household cat faces between tormenting a mouse and eating it. A cat can chase a mouse, pin it down, bite it, and swat it around, but there's one thing a cat can't do, even on YouTube®: use its forelimb to grasp a mouse by the tail and gently evict it from the house. In the next chapter, we'll explain when—and in what kind of animal—such handy skills evolved.

References

1. Qureshi, I. A., Korya, D., Kassar, D., & Moussavi, M. Case report: 84 year-old woman with alien hand syndrome. *Faculty of 1000 Research* 5, 1564 (2016).
2. Ruben, J. The evolution of endothermy in mammals and birds: from physiology to fossils. *Annual Review of Physiology* 57, 69–95 (1995).
3. Rowe, T. B., Macrini, T. E., & Luo, Z. X. Fossil evidence on origin of the mammalian brain. *Science* 332, 955–957 (2011).
4. Long, A., Bloch, J. I., & Silcox, M. T. Quantification of neocortical ratios in stem primates. *American Journal of Physical Anthropology* 157, 363–373 (2015).
5. Killcross, S. & Coutureau, E. Coordination of actions and habits in the medial prefrontal cortex of rats. *Cerebral Cortex* 13, 400–408 (2003).
6. Rich, E. L. & Shapiro, M. L. Prelimbic/infralimbic inactivation impairs memory for multiple task switches, but not flexible selection of familiar tasks. *Journal of Neuroscience* 27, 4747–4755 (2007).
7. Young, J. J. & Shapiro, M. L. Double dissociation and hierarchical organization of strategy switches and reversals in the rat PFC. *Behavioral Neuroscience* 123, 1028–1035 (2009).
8. Guise, K. G. & Shapiro, M. L. Medial prefrontal cortex reduces memory interference by modifying hippocampal encoding. *Neuron* 94, 183–192 (2017).
9. Dias, R. & Aggleton, J. P. Effects of selective excitotoxic prefrontal lesions on acquisition of nonmatching- and matching-to-place in the T-maze in the rat: differential involvement of the prelimbic-infralimbic and anterior cingulate cortices in providing behavioural flexibility. *European Journal of Neuroscience* 12, 4457–4466 (2000).
10. Spellman, T. et al. Hippocampal-prefrontal input supports spatial encoding in working memory. *Nature* 522, 309–314 (2015).
11. Navawongse, R. & Eichenbaum, H. Distinct pathways for rule-based retrieval and spatial mapping of memory representations in hippocampal neurons. *Journal of Neuroscience* 33, 1002–1013 (2013).
12. Backus, A. R., Schoffelen, J. M., Szebenyi, S., Hanslmayr, S., & Doeller, C. F. Hippocampal-prefrontal theta oscillations support memory integration. *Current Biology* 26, 450–457 (2016).
13. Walton, M. E., Bannerman, D. M., & Rushworth, M. F. S. The role of rat medial frontal cortex in effort-based decision making. *Journal of Neuroscience.* 22, 10996–11003 (2002).
14. Rudebeck, P. H., Walton, M. E., Smyth, A. N., Bannerman, D. M., & Rushworth, M. F. Separate neural pathways process different decision costs. *Nature Neuroscience* 9, 1161–1168 (2006).
15. Shenhav, A., Botvinick, M. M., & Cohen, J. D. The expected value of control: an integrative theory of anterior cingulate cortex function. *Neuron* 79, 217–240 (2013).

16. Winstanley, C. A., Theobald, D. E. H., Cardinal, R. N., & Robbins, T. W. Contrasting roles of basolateral amygdala and orbitofrontal cortex in impulsive choice. *Journal of Neuroscience* **24**, 4718–4722 (2004).
17. Kesner, R. P. & Gilbert, P. E. The role of the agranular insular cortex in anticipation of reward contrast. *Neurobiology of Learning and Memory* **88**, 82–86 (2007).
18. Pickens, C. L., Saddoris, M. P., Gallagher, M., & Holland, P. C. Orbitofrontal lesions impair use of cue-outcome associations in a devaluation task. *Behavioral Neuroscience* **119**, 317–322 (2005).

6

Primates of the past

Arboreal achievements

He sat alone in the darkness, gazing at the dying fire, and seeing faces in it. The last face was so horrible and so simian that he gazed at it in amazement. It got so vivid that, with a little uneasy laugh, he felt on the table for a glass containing a little water to throw over it. His hand grasped the monkey's paw, and with a little shiver he wiped his hand on his coat and went up to bed.

—*The Monkey's Paw* by W. W. Jacobs

The Monkey's Paw

In *The Monkey's Paw*, a short story by W. W. Jacobs, a family friend returns to early 20th-century England after two decades in India. He gives the family an enchanted, mummified monkey's paw that has the power to grant its owner three wishes. Stories with this motif nearly always end badly, and *The Monkey's Paw* is no exception. To make a short story even shorter, the husband and wife first wish for enough cash to pay off their mortgage. They get it, but only as compensation for the death of their son. Machinery at his workplace chews him up—fatally and gruesomely. Later, after the funeral and associated falderal, the wife demands that her husband make a second wish. With appalling judgment and insufficient attention to detail, the man uses the monkey's paw to bring their son back to life. As something mangled and awful-beyond-words claws at the front door, the husband manages to wish it away—the final wish—just before the door opens.

We share this story because of the word "paw" in its title. Anyone who has been grabbed by a monkey—as we have on numerous occasions—knows that they have hands, not paws. And monkeys have hands because of the way early primates adapted to life in the trees. As they did, our primate ancestors developed a hindlimb-dominated form of locomotion that freed their forelimbs for grasping and manipulating items in a cluttered world. In Figure 6.1, we illustrate the difference between paws and hands.

The Evolutionary Road to Human Memory. Elisabeth A. Murray, Steven P. Wise, Mary K. L. Baldwin, and Kim S. Graham, Oxford University Press (2020). © Oxford University Press.
DOI: 10.1093/oso/9780198828051.001.0001

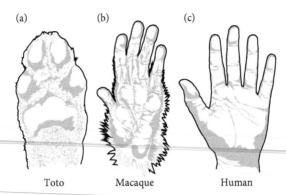

Fig. 6.1 A paw and two hands. Macaques are Old World monkeys.

Part b: Adapted from Dylan F. Cooke, Adam B. Goldring, Mary K. L. Baldwin, Gregg H. Recanzone, Arnold Chen, Tingrui Pan, Scott I. Simon, and Leah Krubitzer, Reversible Deactivation of Higher-Order Posterior Parietal Areas. I. Alterations of Receptive Field Characteristics in Early Stages of Neocortical Processing, *Journal of Neurophysiology*, 112 (10), pp. 2529–2544, Figure 3d, doi.org/10.1152/jn.00140.2014 ©The American Physiological Society (APS). All rights reserved.

Knowing and doing

A discussion of hands and grabbing might seem out of place in a book on memory. You can reach for things with great agility, but you don't know how you do it. You know *that* you reach, and you know *why* you reach, but you don't know *how* you reach. The same goes for all skilled actions. When you throw a dart toward the bullseye, you hope it goes there, but you have no idea how your brain does it. Because skills improve with practice, they must involve memories of some sort, but you remain oblivious to them.

In Chapter 1, we mentioned two forms of memory: cultural and participatory. Together, they compose what we call personal memory, which most scientists call either explicit or declarative memory. These are not the kinds of memory that underlie skills, but skills require memories all the same. Although you remain utterly unaware of the memories that guide your movements, your brain stores the learned information you need for all your skills, including typing, dancing, tool use, and athletics.*

The term "muscle memory" arises from the mysteries surrounding skilled movements. It implies that the human brain outsources the job of remembering how to do things. This isn't entirely out of the question because some animal brains do something like that. The octopus brain, for example, cedes a fair amount of autonomy to each of the animal's eight legs.[1] Each leg uses its own sensations to guide

---

* Some scientists refer to these memories as *procedural*, but this term covers only a fraction of the stored information that falls outside the bounds of personal (explicit) memory.

its actions, so an octopus can do several things at once: multitasking at its best. Vertebrates, however, consolidate the control of skilled actions entirely within the brain. The impression that your muscles have memories of their own comes from the feeling that something "impersonal," something other than your subjective "self," guides your most skillful actions. In Chapter 10, we'll discuss why some memories seem personal and others don't. Here, we'll concentrate on the "impersonal" memories that guide your movements, beginning with the fascinating case of patient D. F.

This patient suffers from a severe impairment in perceiving objects and shapes.[2] Yet she can use vision to guide her hand with remarkable accuracy. When she reaches toward a flat disk, she can move her fingertips directly to its edge, precisely accounting for the disk's exact size and shape. Despite this ability, she cannot say which of two disks is larger. Likewise, she can adjust her hand to move it through a rectangular slot but cannot use her hand to indicate a slot's orientation to someone else. Although she can tell people about crude aspects of her visual world, she can't appreciate its finer points. As usual in such cases, this patient suffers from damage to her cerebral cortex, mainly the temporal lobe.

D. F. has a disorder called *visual agnosia*. This label refers to the fact that she doesn't *know* what she sees (agnosia), but it omits what she can *do* with what she sees, which is a lot. She is a decent doer but a poor perceiver. In Shakespeare's *King Richard III*, the king hires hit men to whack his brother. They assure the king of their dependability by saying:

Talkers are no good doers. Be assured
We go to use our hands and not our tongues.

The king's henchmen might be horrible humans, but by using their hands they are proper primates.

Damage to the parietal lobe causes a different impairment, called *optic ataxia*.* One such patient, C. F., has difficulty reaching to items accurately.[3] In clinical testing, he has trouble orienting a key correctly so that it goes into a padlock. He also points inaccurately to targets in front of him, especially when he's looking at the examiner's nose instead of his target. Nevertheless, he can orient his hand to indicate a slot's orientation to someone else, and he can appreciate the finer details of objects.[4]

These two disorders reveal something important about primate brains. Accurate reaching and hand movements rely, in part, on cortical areas in the parietal lobe, and by comparing the brains of many mammals we know that most of these areas evolved in primates.[5] Patient C. F. has damage to these areas, so he has trouble

---

* Ataxia comes from the Greek word *taxis*, meaning order. By adding a negating prefix, *a*, optic ataxia refers to a disordered form of visually guided movement.

reaching and orienting his hand; patient D. F. has damage to other cortical areas, so she makes these movements with ease.

Nailing life in the trees

As we explained in Chapter 2, early primates adapted to a nocturnal life confined to trees. With forward-facing eyes, grasping hands and feet, fingernails instead of claws, and a set of new brain areas, early primates became specialists in using vision to reach for, grasp, and handle items in their arboreal world. By bringing both eyes to bear on an item, forward-facing eyes improved their perception of distances and contours. As another adaptation to arboreal life, early primates used their legs to launch themselves from branch to branch, and when they reached their destination, they grasped a new perch with their feet and hands: a *leaping–grasping* mode of locomotion. With four grasping limbs, they could spare one hand to grab items, as well as to bring food to the mouth.

To perform these feats, early primates had to solve several problems. For example, they needed to account for movements of their body as flimsy branches bounced up and down. They also had to consider the configuration of their arm every time they reached for something. In their unstable, swaying world, their arm rarely began a movement the same way twice. In addition, these animals had to maneuver their hands around branches that blocked a direct route to whatever they wanted. And once they grasped a bit of food, they had to avoid squeezing it so hard that it disintegrated; yet, they needed to grasp it firmly enough to detach it from the tree. Insects, too, could be squashed by excessive force, but insufficient force might enable them to escape. Early primates could avoid damaging their food by pulling a branch toward their mouth and gnawing off the desired item. But this approach also required tremendous (or, perhaps, "tree-mendous") agility. In short, dining in the trees was no picnic for our early primate ancestors.

New cortical areas in the parietal, temporal, and frontal lobes helped early primates surmount these challenges. In Figure 1.2, we illustrate the frontal lobe of human brains. In addition to elaborations of the parietal and temporal lobes, new *premotor* areas appeared near the back of the frontal lobe, along with new *prefrontal* areas in the front.[6]

- Premotor and parietal areas specialize in neural representations that guide reaching and grasping movements.[7] These areas represent the locations of items in the visual world, use memories to translate these locations into a precise pattern of muscle-generated forces, and adjust these memories to reduce future errors. The next section takes up this topic.
- One new prefrontal area interacts with the temporal lobe and the amygdala to update the subjective valuation of objects and actions, as we'll explain later.

- Another new prefrontal area interacts with the temporal and parietal lobes to help primates search for and attend to valuable items, a process we'll discuss briefly near the end of this chapter.

Reaching for the sights

Arrows of action

In Figure 6.2, we illustrate a bushbaby (also known as a galago), a small primate related to lemurs. The dashed line indicates where the animal concentrates its attention. In this case, the bushbaby looks intently at the eyes of the Wicked Witch of the West to detect her intentions. It makes sense for the bushbaby to keep a close eye on this dangerous sorcerer, but it can't forego food forever. It wants to eat a persimmon, but it also needs to avoid capture by the witch, who surely wants a bushbaby to supplement her brigade of flying monkeys. In this situation, the parietal cortex uses sensory information from vision to represent the location of both a goal (the persimmon) and the bushbaby's hand. In Figure 6.2, we depict these locations with arrows (vectors), both of which begin at the attended location, the witch's eyes, and point somewhere else. The *goal vector* points from the witch's eyes to the bushbaby's goal, the persimmon; the *hand vector* points from the witch's eyes to the bushbaby's hand. Studies of neuronal activity indicate that the parietal cortex represents hand and goal locations in just this way.

A third arrow results from the subtraction of one of these vectors from the other, a computation that generates a movement plan. To see why subtraction produces a movement plan, imagine the tail of the goal vector, its red part, moving along the hand vector from the attended location (the witch's eyes) until it reaches the hand's location. The process of moving the hand vector's tail in this way is equivalent to subtracting it from the goal vector, and what remains is called the *difference vector*. After many experiments, scientists have a pretty good idea about how neural networks in the premotor and parietal cortex compute and represent difference vectors.[8]

Analysis of these neural representations has kept a bunch of scientists busy for decades, but a lot of this neural algebra seems unnecessary at a first glance. Why not simply represent a movement plan—the difference vector—directly and dispense with the goal and hand vectors? Some animals might do this,* but we know that the primate brain doesn't work that way because whenever the eyes move, the brain recalculates all three of the vectors in Figure 6.2.[9] You can appreciate the reason for

---

* The neural mechanisms of reaching are much better understood in primates than in other mammals. There is, however, no evidence that mammals such as rats, mice, raccoons, or cats plan their movements in an extrinsic coordinate frame, as primates do.

Fig. 6.2 Arrows of action. The bushbaby's goal is to obtain a persimmon. Arrows stand-in for three neural representations: of the goal's location; of the location of a bushbaby's hand; and of the difference between them, which corresponds to a motor plan. The bushbaby attends intently on the witch's eyes, as indicated by the dashed line.

these recalculations by concentrating on the tail of both the goal and hand vectors in Figure 6.2. Both are tied to the precise place that the bushbaby directs its attention. So, despite the fact that neither the hand nor the target moves, the hand and goal vectors change whenever the bushbaby shifts its attention. The reason for all this complexity is that the primate brain evolved to represent locations in terms of the visual world. Recall from Chapter 5 that rodents can navigate through a maze in two ways: one based on the outside world (go east or west, as in Fig. 5.3a); the other based on its own body (go right or left, as in Fig. 5.3b). The same idea applies to reaching in primates. Remarkably, the primate brain has evolved to guide reaching based on vision of the outside world (an extrinsic coordinate frame) rather than relative to its own body (an intrinsic coordinate frame), a property that reflects the dominance of vision in the life of primates.

Because of all this, the fact that the Wicked Witch flies by the bushbaby's branch doesn't matter as much as it might. The bushbaby can monitor the witch as she flies by, ensuring that she doesn't swerve menacingly, all the while planning to grab a juicy piece of fruit. The bushbaby's hand can move quickly along the *difference vector* without the need to look at the fruit until the very last moment. Because your primate ancestors evolved this way of reaching, you can take advantage of your exquisite vision to represent a motor plan, but you don't need vision to execute that plan. For example, if you remember where you saw a cup of coffee, you can reach for it while attending to something else. Likewise, the bushbaby can keep a watchful eye on the Wicked Witch of the West even as it plans to grasp a persimmon. And, after grabbing the fruit, the bushbaby can resume its witch watch, prepared to launch itself to safety at the slightest sign of bad intent.

Before the bushbaby gets the persimmon, however, the brain needs to make one additional calculation. Premotor areas of the cortex need to transform the *difference vector* into a pattern of forces, which muscles generate to make the planned movement. These calculations require a form of memory called an *internal model*,[8] which takes into account the properties of the arm and hand and the forces needed to detach a fruit from its stem (without crushing it), based on past experience. Internal models resemble the memories needed for skills like playing the piano or typing in two ways: they exist in our brains without our becoming aware of them; and they provide us with new capabilities. A variety of factors, including arm orientation and body posture, influence the pattern and magnitude of muscle forces needed to obtain a piece of fruit. So, small errors are commonplace. After the bushbaby gets the persimmon, its brain records any inaccuracies in force generation so that, with practice, it can reduce these errors in the future.

An autopilot within

Not only can the brain reduce errors from one reaching movement to the next, but it can also adjust movements on the fly. Scientists have likened this mechanism to autopilot control in aircraft.

In one experiment, the target of a reaching movement jumped abruptly after people had started to reach for it. The target was a small spot of light, which appeared suddenly before it jumped just as suddenly to a new location. So, naturally, the participants shifted their vision to look at the spot at the same time as they started to reach toward it. In every case, after the target jumped, the trajectory of their hands quickly and smoothly adjusted in midflight, ending accurately on the target. Because the spot had jumped while they shifted their vision, no one ever noticed that the target had moved. When asked, the participants denied that the spot had moved, despite the fact that they had just reached to a different place than originally intended.[10] Even when the experimenters asked them to reach to the target's original location, they couldn't resist making the adjustment.

Research has shown that the parietal cortex contributes to this kind of autopilot control. By placing an electronically controlled magnet over a person's scalp, scientists can generate electrical current in the cortex underneath. When researchers used this method to disrupt the function of the parietal cortex, people reached to the target's original location instead of its new one, and patients with damage to their parietal cortex do the same thing.[11, 12]

You might wonder what autopilot control has to do with memory. After all, a target was visible at all times during these experiments; no one needed to remember its location. But memories are required to translate visual locations into the neural signals that control movements. Put another way, the brain stores the memories needed to convert a visually sensed location into an appropriate pattern of muscle activity. Experts refer to these memories as visuomotor transforms and internal models, and they don't often use the word memory while discussing them. For that reason, the fruits of motor learning are often neglected in general discussions of memory. Like other forms of memory, however, these kinds depend on learning: in this case from previous reaching movements and any errors that occurred. For reaching movements, your brain computes visuomotor transforms and implements internal models covertly, which produces a sense of impersonal, autopilot control.

Action representations

A different form of cortical stimulation also reveals something about the memories that guide movements. Experimenters can use fine wires to stimulate small patches of cortex with an electrical jolt. When sustained for several seconds, this stimulation evokes coherent movements, which often resemble those made during the natural behavior of primates, including grasping, reaching, groping, climbing, protective, defensive, or feeding movements.[13] Although the interpretation of these observations remains somewhat controversial, it seems likely that these movements reflect stored representations—memories—arising from prior experience in performing these actions.

An example from monkeys is particularly instructive. When experimenters stimulated a premotor area called the ventral premotor cortex, monkeys made coordinated movements of their arm, hand, head, mouth, and lips, much like they did when they brought food items to their mouth.[13] In a related experiment, inactivation of the same cortical area caused inaccuracies in grasping an object.[14] Taken together, these results indicate that this area stores some of the memories that guide coordinated feeding movements. This makes sense because, in addition to this new cortical area, early primates had several evolutionary innovations related

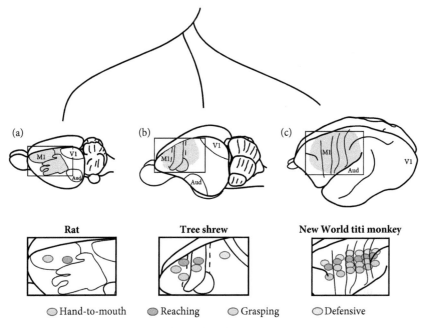

Fig. 6.3 Action representations. The results of cortical-stimulation experiments in rats (a), tree shrews (b), and titi monkeys (c). Each colored circle or oval indicates an action evoked by electrical stimulation of that small patch of cortex. The abbreviation M1 stands for the primary motor area; V1 stands for the primary visual area; and Aud refers to a group of auditory areas.

Parts a and b: Adapted from Baldwin MKL, Cooke DF, and Krubitzer L, Intracortical Microstimulation Maps of Motor, Somatosensory, and Posterior Parietal Cortex in Tree Shrews (*Tupaia belangeri*) Reveal Complex Movement Representations, *Cerebral Cortex*, 27 (2), pp. 1–18, Figure 1, doi: 10.1093/cercor/bhv329 © The Author 2016. Part c: From Baldwin MKL, Krubitzer L, Parallel Motor Pathways in the Neocortex of Tree Shrews and Monkeys. Program No. 310.13. 2018 Neuroscience Meeting Planner. San Diego, CA: Society for Neuroscience, 2018. Online.

to feeding, including hands instead of paws and a hindlimb-dominated form of locomotion that freed the hands for reaching and manipulating food items. And many primate species have similar representations because they will have made similar movements during their lives.

In Figure 6.3, we illustrate the results of this kind of experiment in a rat, a tree shrew, and a New World primate called titi monkeys.[15,16] Tree shrews are among the closest living relatives of modern primates, with rodents such as rats being less closely related.* The colored shapes in Figure 6.3 indicate the kinds of movement evoked by prolonged electrical stimulation at various sites. Rats have only a little

---

* In Figure 2.4, we illustrate these relationships on the evolutionary tree of this group of mammals.

parietal or premotor cortex, so one of the few areas large enough to stimulate is called the primary motor cortex. This cortical area probably evolved in early placental mammals, and it provides these mammals with their most direct cortical control of body movements. Stimulation of the primary motor cortex evokes a few coherent actions in rats, reaching and pawing (Fig. 6.3a), but not much else. In tree shrews, coherent actions follow stimulation of both the parietal cortex and the primary motor cortex, including reaching, grasping, and hand-to-mouth movements (Fig. 6.3b). In titi monkeys, stimulating the premotor, primary motor, or parietal cortex evokes a wide range of actions (Fig. 6.3c). These studies suggest that the basic pattern of organization seen in monkeys descended from early primates, although the last common ancestor of tree shrews and primates probably had a nascent form of what ultimately evolved in primates.

Patches of both the parietal and frontal cortex represent the same action, and these patches tend to send connections to each other. They probably work together in representing coherent actions.[13,17] This finding is not particularly surprising in light of the idea that most of these premotor and parietal areas evolved at more-or-less the same time.

Affordances

The term *affordance* refers to an action associated with an object or surface. A staircase, for example, elicits memories of ascending or descending a surface shaped like that, and a baseball practically "wants" to be grasped and thrown. To an infant, however, staircases and baseballs lack such affordances, at least initially. Like other memories, affordances come from experience, and they exert a powerful influence over our actions. One reason is that the size, shape, and weight of an object constrain what can be done with it (or to it).

In humans, damage to the parietal cortex impairs affordance memories.* As a result, patients with such damage forget how to use objects, such as tools, including those they used adeptly in the past. Although these patients have normal strength and can make simple movements, they can't use a pair of scissors or a hammer, despite knowing their names. (In Chapter 9,† we'll discuss patients who have a problem remembering the names of tools because of damage to different brain areas.)

As we discussed throughout Chapter 5, representations compete with each other to control behavior, and this idea also applies to affordances. Viewed as an object, an iPad can be used to check on any e-mail messages that might have arrived in the past five minutes, but it can also be used to smash a fly. As we'll discuss

---

* As we said in the section entitled "Known and doing," patient C. F. also has a damaged parietal cortex, and he has a similar impairment. He can't use vision to guide his movements accurately.
† In the section entitled "Four-legged ducks."

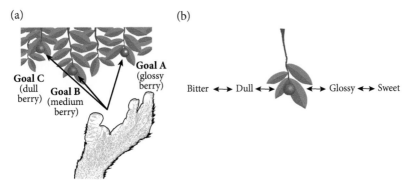

Fig. 6.4 Berry picking. (a) Arrows indicate the distance and direction of three berries from a monkey's hand. (b) Arrows indicate the association of a glossy berry with a sweet taste and a dull berry with a bitter taste.

next, our new prefrontal areas guide such choices by biasing a competition among affordance memories stored by parietal and premotor areas.

Deciding on desirability

Imagine deciding what kind of fruit to buy before hitting the road to a market. You want apples, so off you go to get some. Now imagine the opposite approach: going to a market and then deciding what kind of fruit to buy. Both approaches have their place, but they differ in terms of when a *valuation* process occurs: before versus after seeing the alternatives. We call these foraging strategies the *desirability-first* and *desirability-second* approaches, respectively.

Paul Cisek[18,19] has developed and tested a model of how the primate brain manages the desirability-second approach. Imagine a monkey choosing among three berries, as illustrated in Figure 6.4a, based on subtle visual cues such as glossiness. According to Cisek, the monkey first establishes three movement plans, each represented by an arrow. A little while later, the brain uses a valuation process to choose the best berry.

An experiment on monkeys provided support for this idea.[18] The monkeys saw two circles, both of which served as a potential target for a reaching movement. A colored border around each circle indicated its value: how much juice the monkey would receive for reaching to it. The activity of neurons in the premotor cortex indicated that this region first represented each of the potential reaching targets, and only later—tens of milliseconds later—did these representations begin to incorporate their relative value. This finding indicates that neurons in the premotor cortex represent a potential movement first and its value second. Given three berries within reach, the best berry will become the pending goal. But what makes one berry seem more valuable than another?

The arrows in Figure 6.4b depict associations between a berry's visual features and its desirability. Let's assume that ripe berries are glossy and sweet; dull berries, although decent in taste and nutritious to a degree, have less sugar and taste a little bitter. The primate temporal lobe has developed sophisticated specializations for representing the finer aspects of an object's appearance, such as its precise color and glossiness, and the orbital cortex has access to these representations through its neural connections. As we explained in Chapter 2,* the largest part of the orbital cortex is unique to primates, as are several parts of the temporal lobe and most premotor areas. According to Cisek's model, the orbital cortex sends a valuation signal to premotor areas, and the relative desirability of the three berries generates a bias among competing movement plans. The most highly valued option receives the strongest bias, so the monkey reaches for and grasps the best berry.

In Cisek's desirability-second model, the brain begins by activating movement representations in premotor and parietal areas, then the orbital cortex chimes in to boost the movement-control signal that will yield the best berry. In contrast, when applying the desirability-first approach, the brain begins by activating the representations of competing items in the orbital cortex, such as an apple and a persimmon. In this way, the most valuable item can be selected before planning a movement to either item, and even before seeing them. Camillo Padoa-Schioppa[20] has studied desirability-first choices, and the results of his experiments point to primate-specific parts of the orbital cortex as the key contributors to this way of choosing things.

In either case, the orbital cortex contributes something important to the valuation process: an assessment of desirability that is based on the updated, moment-to-moment needs of an individual. As we explained in Chapter 3,† the *devaluation task* can be used to study updated valuations.[21] This task takes advantage of the fact that, like eating a gallon of ice cream or an enormous bowl of popcorn, enough is enough and eventually too much. After pigging out on one kind of food, the sight of any more of it becomes unpalatable.

A peculiar weight-loss program from the 1980s depended on this disagreeable experience. Every day, dieters could eat as much of one food as they wanted: chunks of pineapple, for example. Unfortunately, the rules of the diet prohibited these poor souls from eating anything else that day. The diet's advocates advertised it with an attractive slogan: "Eat as much as you want and still lose weight!" But it made most people so miserable that, in the end, they abandoned the diet fairly quickly.

In an experiment related to this phenomenon, monkeys first learned to choose a particular object, which always covered one kind of food. The example in

---

* The section entitled "Primates go out on a limb."
† In the section entitled "Instrumental memories."

Figure 6.5a illustrates a purple dome that covered a sweet, red berry and a green cube that covered an empty food well. On other trials, which are not illustrated, an orange cone covered a peanut and a different object covered an empty food well. In both cases, the monkeys learned to choose the object that covered food. The monkeys later faced a memory test, which we illustrate in Figure 6.5c. To get a berry, which they preferred over a peanut, the monkeys needed to remember that choosing the purple dome uncovered a berry and that choosing the orange cone uncovered a peanut.

In the next phase of the experiment, which we depict in Figure 6.5b, the monkeys ate the berries until they couldn't consume any more. Each eaten berry devalued the next one until the monkeys reached a state of satiety. Scientists call this state *selective satiation* because it applies mostly to one kind of food. The monkeys then faced the same choice as before: the purple dome versus the orange cone. Normal monkeys usually switched their choice from the purple dome to the orange

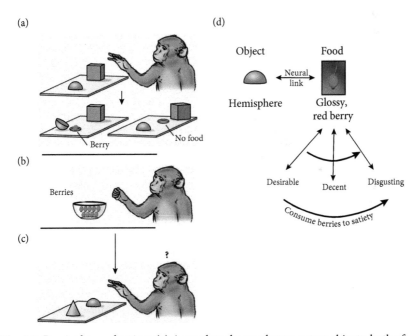

Fig. 6.5 Berries beyond satiety. (a) A monkey chooses between two objects, both of which cover a foodwell. One well contains a berry; the other is empty. (b) The monkey then eats as many berries as it wants. (c) Later, the monkey faces a choice between two objects. One object is associated with berries; the other with peanuts. (d) As a monkey consumes a bellyful of berries, their desirability decreases to the point that they become utterly revolting.

Adapted from Elisabeth A. Murray and Sarah E. V. Rhodes, 'Monkeys without an Amygdala,' in David G. Amaral and Ralph Adolphs, ed., *Living without an Amygdala*, pp. 252–275, Figure 9.1 © 2016 The Guilford Press.

cone, so they got a peanut. After a surgical removal of their orbital cortex, however, monkeys didn't make this shift very often. They continued to choose the purple dome, thereby obtaining the devalued and now-undesirable berry. Similar results have been obtained when monkeys chose between two different actions instead of between two different objects.[21]

In Figure 6.5d, we imagine how this shift occurs. In addition to representing a berry's color, glossiness, smell, and taste, neurons in the orbital cortex also represent its moment-to-moment desirability. As an animal consumes a big bowl of berries, their value declines from "highly desirable" to "decent" and then—after yet more gluttony—to "disgusting." At some point, the desirability of a peanut will become greater than the desirability of a berry, and a monkey will shift its choice to obtain a "decent" peanut instead of a now-disgusting berry.

For this shift to occur, the orbital cortex needs to have access to neural representations of a food's updated desirability as well as the visual properties of the purple dome and the glossy, red berries. Fortunately, the orbital cortex has just the connections to do this job. It has access to smell and taste representations in nearby areas of orbital cortex—the areas shared with rodents and other mammals—and it has access to color, shape, and glossiness representations in the temporal lobe. Crucially, it also has connections with the amygdala, which provides an up-to-the-minute desirability rating for food items. The primate-specific parts of the orbital cortex are therefore in a unique position: they can integrate all of these subsidiary representations into a single, more-complex one. Accordingly, only these areas can link the updated desirability valuation of a food item with the visual features of objects, in this case the purple dome and the orange cone.

The indirect nature of these choices deserves a comment. It is one thing to choose between a berry and a peanut, another to choose between two inedible objects that hide food items. This distinction highlights the fact that primates often make choices based on inedible things in their visual world, which often hide the good stuff. The outer covering of a fruit or nut reveals a lot about what is on the inside, even if the outside is worthless (or worth less). After all, who eats peanut shells? A termite mound has no nutritional value, but the termites inside are—the way many primates see it, at least—highly desirable. In laboratory experiments, the ability to make choices based on external appearances empowers primates to choose among objects that are associated with food items.

Searching for value

In a cluttered world of leaves and branches, it can be challenging to find and keep track of valuable items. Another early primate part of the prefrontal cortex, called the *frontal eye field*, uses its interactions with both the temporal and parietal lobes to help primates search for and attend to things. Areas in the temporal lobe represent *qualitative* aspects of vision, such as shape, color, and glossiness; the back part of the parietal lobe represents *quantitative* aspects of vision, such as the number of items.[22]

Representations in the frontal eye field guide the search for items based on both qualitative and quantitative information. As we illustrate in Figure 6.4a, primates often choose among similar items, such as berries that differ mainly in glossiness. The frontal eye field can enhance the detection of glossy berries by sending signals to the temporal lobe that favor representations of glossy redness at the expense of other representations, including dull redness: a process called *top-down attention*.[23] * Primates can use a similar mechanism to search for a large cluster of berries, as opposed to a small cluster.

By deploying top-down attention for both qualitative and quantitative features of the visual world, the frontal eye field can favor cortical representations of the largest number of sweet, glossy berries. When the best berries are dispersed within a clutter of leaves, branches, and barely edible berries, this process provides an important evolutionary advantage. Importantly, it didn't come completely out of the blue. The frontal eye field built on the function of older prefrontal areas, the ones we discussed in Chapter 5. There we said that when the first prefrontal areas emerged, in early mammals, they also exerted a top-down bias: on memories that complete to control behavior. The frontal eye field extended this function to the competition among sensory representations.

Primate innovations reconsidered: a far-reaching legacy

Early primates evolved a suite of new traits—of both brain and body—as they adapted to a life confined to trees. A list of these traits reads something like a description of ourselves: forward-facing eyes; grasping hands with fingernails; vision-dominated perception; movements guided primarily by vision; locomotion that depends mostly on the legs; a long life; and a large brain.

Within this large brain, new cortical areas contributed to solving a set of interrelated problems: how to use vision to reach accurately while perched on swaying branches; how to generate the precise amount of force needed to grasp and hold valuable items; how to find the best food items among the clutter of branches and leaves; and how to choose among objects and actions based on updated, moment-to-moment needs.

In Figure 6.6, we use green labels to indicate new representations that emerged in the evolutionary transition to true primates. New representations in the parietal and premotor areas empowered early primates to reach for and grasp things in

---

* We mentioned top-down attention in Chapter 3's section entitled "Instrumental memories."

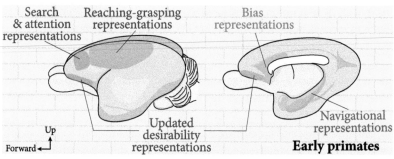

Fig. 6.6 Some primate augmentations. Green labels and shading indicate some of the representations that emerged in early primates, placed on a brain drawing based on a fossil primate that lived 35–55 million years ago. A side view of the brain is at the lower left; a view of half the brain from the middle (with the other half removed) is at the lower right. Representations inherited from ancestral vertebrates and early mammals are indicated with blue and pink labels, respectively. Just as the Tin Man joins the Scarecrow and Dorothy on their journey along the yellow brick road, several new kinds of representations (green) joined older ones (blue and pink) as primate evolution marched along. In the background, we depict icons from the past: a compass device marks the beginning of the evolutionary road to human memory; and the cornfield commemorates Dorothy's encounter with the Scarecrow near the dawn of mammals.

their fine-branch world and to do so with phenomenal agility. Updated desirability valuations in the orbital cortex improved foraging choices based on vision; and top-down attention driven by representations in the newly evolved frontal eye field helped them search for and attend to valuable items. With these advantages, our early primate ancestors could thrive in the trees, safe from predators below.

Earlier, we mentioned going to a market for apples. In *The Wizard of Oz*, when Dorothy picks an apple off of a tree, the tree grabs it back and slaps her on the hand. A startled Dorothy then endures the tree's tongue-lashing:

> TREE: What do you think you're doing?
> DOROTHY: We've been walking a long ways, and I was hungry . . . Did you say something? . . .
> TREE: Well, how would you like to have someone come along and pick something off of *you*?

Back in Kansas, trees don't grab things. They have plenty of limbs, but no hands. As part our inheritance from early primates, a large part of our brain consists of cortical areas that control our limbs and hands. But the memories they store seem alien: "out of reach" in a metaphorical sense. The word *grasping* is commonly used as a metaphor for understanding, but an understanding of grasping (in a literal sense) is beyond our grasp (in a metaphorical sense), at least without some highly specialized science and mathematics. We have no subjective appreciation of the memories that we use for grasping, so they seem "impersonal." In Chapter 10, we'll present an idea about what makes other memories seem personal.

In the epigraph of Chapter 4, we quoted something that Hunk said to Dorothy: "don't go by Miss Gulch's place," he tells her, "and you won't get in no trouble. See." In the next chapter, we'll explore why Hunk says "see" when he refers to understanding.

References

1. Sumbre, G., Gutfreund, Y., Fiorito, G., Flash, T., & Hochner, B. Control of octopus arm extension by a peripheral motor program. *Science* **293**, 1845–1848 (2001).
2. Milner, A. D. & Goodale, M. A. *The Visual Brain in Action* (Oxford, UK: Oxford University Press, 2007).
3. Rondot, P., De Recondo, J., & Dumas, J. L. Visuomotor ataxia. *Brain* **100**, 355–376 (1977).
4. Ambron, E., Lingnau, A., Lunardelli, A., Pesavento, V., & Rumiati, R. I. The effect of goals and vision on movements: a case study of optic ataxia and limb apraxia. *Brain and Cognition* **95**, 77–89 (2015).
5. Padberg, J. et al. Parallel evolution of cortical areas involved in skilled hand use. *Journal of Neuroscience* **27**, 10106–10115 (2007).

6. Preuss, T. M. & Goldman-Rakic, P. S. Myelo- and cytoarchitecture of the granular frontal cortex and surrounding regions in the strepsirhine primate *Galago* and the anthropoid primate *Macaca. Journal of Comparative Neurology* **310**, 429–474 (1991).

7. Borra, E., Gerbella, M., Rozzi, S., & Luppino, G. The macaque lateral grasping network: a neural substrate for generating purposeful hand actions. *Neuroscience and Biobehavioral Reviews* **75**, 65–90 (2017).

8. Shadmehr, R. & Wise, S. P. *The Computational Neurobiology of Reaching and Pointing: A Foundation for Motor Learning* (Cambridge, MA: MIT Press, 2005).

9. Colby, C. L., Duhamel, J. R., & Goldberg, M. E. Visual, presaccadic, and cognitive activation of single neurons in monkey lateral intraparietal area. *Journal of Neurophysiology* **76**, 2841–2852 (1996).

10. Day, B. L. & Lyon, I. N. Voluntary modification of automatic arm movements evoked by motion of a visual target. *Experimental Brain Research* **130**, 159–168 (2000).

11. Prablanc, C., Desmurget, M., & Grea, H. Neural control of on-line guidance of hand reaching movements. *Progress in Brain Research* **142**, 155–170 (2003).

12. Grea, H. et al. A lesion of the posterior parietal cortex disrupts on-line adjustments during aiming movements. *Neuropsychologia.* **40**, 2471–2480 (2002).

13. Graziano, M. S., Taylor, C. S., & Moore, T. Complex movements evoked by microstimulation of precentral cortex. *Neuron* **34**, 841–851 (2002).

14. Fogassi, L. et al. Cortical mechanism for the visual guidance of hand grasping movements in the monkey: a reversible inactivation study. *Brain* **124**, 571–586 (2001).

15. Baldwin, M. K., Cooke, D. F., & Krubitzer, L. Intracortical microstimulation maps of motor, somatosensory, and posterior parietal cortex in tree shrews (*Tupaia belangeri*) reveal complex movement representations. *Cerebral Cortex* **27**, 1439–1456 (2017).

16. Baldwin, M. K., Halley, A. C., & Krubitzer, L. A. The functional organization of movement maps in New World titi monkeys. *Annual Meeting of the Society for Neuroscience,* Program number 316.06 (2017).

17. Stepniewska, I., Gharbawie, O. A., Burish, M. J., & Kaas, J. H. Effects of muscimol inactivations of functional domains in motor, premotor, and posterior parietal cortex on complex movements evoked by electrical stimulation. *Journal of Neurophysiology* **111**, 1100–1119 (2014).

18. Pastor-Bernier, A. & Cisek, P. Neural correlates of biased competition in premotor cortex. *Journal of Neuroscience* **31**, 7083–7088 (2011).

19. Cisek, P. Making decisions through a distributed consensus. *Current Opinion in Neurobiology* **22**, 927–936 (2012).

20. Padoa-Schioppa, C. Neurobiology of economic choice: a good-based model. *Annual Reviews in Neuroscience* **34**, 333–359 (2011).

21. Murray, E. A. & Rhodes, S. E. V. in *Living Without an Amygdala* (eds D. G. Amaral & R. Adolphs), pp. 252–275 (New York, NY: Gilford, 2015).

22. Brannon, E. M. The representation of numerical magnitude. *Current Opinion in Neurobiology* **16**, 222–229 (2006).

23. Desimone, R. & Duncan, J. Neural mechanisms of selective visual attention. *Annual Reviews in Neuroscience* **18**, 193–222 (1995).

7

Anthropoid adaptations

Seeing scenes and signs

We're off to see the Wizard.
The Wonderful Wizard of Oz
We hear he is a whiz of a wiz
If ever a wiz there was . . .

When Dorothy starts down the yellow brick road, her intentions extend well be-yond merely *seeing* the Wizard: wonderful, whizzy, or otherwise. Vision so dom-inates human cognition that *seeing* means much more than viewing the world. We try to *see* what the problem might be; we *see* to it that something is done; we *see* someone to the door; and—when things get very bad—we *see* a lawyer. Dorothy doesn't go to the Emerald City simply to *see* the wonderful Wizard; she intends to ask the whiz of a wiz for a way back to Kansas. And as an expert on hot air, the shameless charlatan knows just how a gasbag can return Dorothy to her no-place-like home.

In this chapter and the next one, we'll explore the world of anthropoid primates and especially their vision. We explained in Chapter 2* that anthropoids—the lin-eage that includes monkeys, humans, and apes—began as small animals. As their descendants grew larger, they traveled long distances to obtain food, usually in daylight. Vision helped them decide where to go and how to avoid danger. Later in this chapter, we'll describe some of the visual innovations that emerged just be-fore and during anthropoid evolution, but first we'll focus on one of the upshots of anthropoid vision: face recognition. In *The Wizard of Oz*, the same actor por-trays five different characters: Professor Marvel; the keeper of the city gates; a guard in the Emerald City; a carriage driver; and the Great and Powerful Wizard of Oz. It's easy for most of us to see that the same actor plays all of these roles, but some people can't.

---

* In the section entitled "Anthropoids arrive."

The Evolutionary Road to Human Memory. Elisabeth A. Murray, Steven P. Wise, Mary K. L. Baldwin, and Kim S. Graham, Oxford University Press (2020). © Oxford University Press.
DOI: 10.1093/oso/9780198828051.001.0001

Distinctions from conjunctions

G. G. is a right-handed male and retired computer engineer born in
1942 . . . [He] complained about severe difficulties to recognize familiar faces: ex-colleagues, neighbours, shopkeepers, etc. He also had
trouble recognizing many famous people on television (actors, politicians, sportsmen, etc.), his colleagues in a painting class he met on a
weekly basis as well as old friends from a war veterans association. He
was also no longer able to follow TV series because he was mixing up
the characters and was unable to learn new faces.[1]

This patient has a disorder called *prosopagnosia*—from the Greek for not knowing
(*agnosia*) faces (*prosópou*). As the name implies, these patients have trouble recognizing and remembering people via face recognition.

Three concepts—*feature conjunction, feature overlap*, and *feature ambiguity*—
help explain prosopagnosia. A good example of a visual *feature* is color; another is
shape. Some visual areas specialize in representing simple combinations of visual
features, such as a green rectangle: a color and a shape bound into a single neural
representation called a *feature conjunction*. Some parts of the temporal lobe harbor
more complex feature conjunctions, meaning that they combine many features
into a single, unique representation, such as the façade of a house containing many
conjunctions and shapes (including our green rectangle—the door). Because one
face shares many features with other faces—we all have noses, after all—it takes
a detailed analysis of differences among many features to tell one face from another. So, we usually identify someone based on their unique combination of facial features. These representations reside in the temporal lobe, so damage to this
brain region—which eliminates or degrades these representations—causes an impairment in distinguishing and remembering faces: prosopagnosia. A patient with
this malady would never notice that Frank Morgan portrays five characters in *The
Wizard of Oz*.

People automatically become experts in recognizing and identifying each
other based on facial feature conjunctions. In contrast, only a few people get
equally good at identifying individuals of other species. A horse is a horse, of
course, of course, but it can be difficult to tell one horse from another. Color
helps. For example, when Dorothy notices a purple horse hauling her through
the Emerald City, she declares: "I've never seen a horse like that before." The carriage driver, played by Frank Morgan, of course, explains why: "There's only one
of him and he's it. He's the Horse of a Different Color you've heard tell about."
Aside from color, horses can be hard to tell apart. A typical dark brown horse
shares so many features with other dark brown horses that "if you've seen one,
you've seen 'em all." In more formal terms, two dark brown horses of the same

The image includes a sign reading: Darwin Derby (D) Road to Fun

Fig. 7.1 A day at the races. A pig wins the Kentucky Derby, with a horse far behind and a mule in between.

size have a high degree of *feature overlap*, which creates *feature ambiguity* that makes them difficult to distinguish from each other. A dark brown horse also shares many features with a dark brown mule of comparable size, but they have a little less feature overlap, in part because of a mule's longer ears. Easier yet is distinguishing a horse from a pig. Even if a pig has a sign on it that says "horse," the sign won't fool anyone because pigs and horses share relatively few features and so have a low level of feature overlap. No matter how devoutly somebody believes a pig to be a horse, no pig will ever win the Kentucky Derby—not even a racing pig. In Figure 7.1, we illustrate this fantasy for a reason that we'll explain in Chapter 9.*

Cortical areas in the temporal lobe empower us to recognize other people by representing unique combinations of facial features, but temporal areas aren't alone in representing feature conjunctions. The way we see it, the entire cerebral

---

* In the section entitled "A horse is a horse."

cortex performs a version of this function, with each area specializing in storing a different combination of features:

- Areas in the inferior temporal lobe* represent conjunctions of *qualitative* features[†] in the visual world, such as the complex combinations of colors, shapes, and surface appearance that make up human faces, objects, and animals;
- Some parietal areas represent conjunctions of *quantitative* features in the visual world, such as numbers, order, distances, directions, and durations;
- Other parietal areas represent combinations of tactile sensations, along with other sensory signals sent from the body to the brain;
- Occipital areas represent simple visual conjunctions and elemental features, such as brightness or color;
- Superior temporal areas* represent acoustic feature conjunctions, such as loudness and pitch, melodies, and speech;
- Frontal areas represent target–value, action–value, and object–value conjunctions, which we discussed in Chapter 6 (for premotor and early-primate prefrontal areas), along with other conjunctions that we'll take up in Chapter 8 (for anthropoid prefrontal areas).

Taken together, feature conjunctions underlie our ability to remember and identify faces, objects, words, and music, among much else.

Eagle-eyed primates

From monochrome to trichromacy

Anthropoid primates are often called "visual animals" despite the fact that nearly all vertebrates have eyes and can see the world, often quite well. The reason is that an ancestor of anthropoids, the founding *haplorhine* primates, evolved the *fovea*, a Latin word that refers to a small "pit" in the retina. We mark this ancestral species with a pink rectangle in Figure 2.4. Like the retina of other vertebrates, the primate retina detects light and color and transmits these sensations to the brain. The fovea is a specialized part of the retina that supports color vision at an extraordinarily fine level of detail. When you say that you're "looking at something," it's usually the fovea that you're looking with, a concept known as *central* (as opposed to *peripheral*) vision. As daytime foragers, foveal vision was especially important to our anthropoid ancestors.

---

* The modifier *inferior* refers to a location near the bottom of the temporal lobe, not to a value judgment about it. Likewise, the *superior* temporal cortex is at the top of the temporal lobe.
[†] In *Vision and Art* (Harry N. Abrams, New York, NY, 2014), Margaret S. Livingstone illuminates the qualitative aspects of vision.

We know that the fovea evolved in the founding haplorhine primates because all of its descendant species have one and no other primates do. All the same, it can be confusing to say "the fovea evolved in primates" because many other animals also have one. In fact, foveas have evolved independently several times in vertebrates.[2] Birds of prey, for example, have more powerful foveas than primates do, and some have two foveas in each eye. The title of this section refers to the renowned vision of eagles, which can detect prey from miles away. Although anthropoids aren't quite "eagle-eyed," they come close, and the primate fovea makes such powerful vision possible. So, it's correct to say that "the fovea evolved in primates," despite the fact that the same thing happened in other animals, too.

Later in anthropoid evolution, an improved system for color vision emerged in Old World primates, a group that includes Old World monkeys, apes, and humans. New World monkeys, such as marmosets and titi monkeys, live in Central and South America. In contrast, Old World primates include the monkeys indigenous to Africa and Asia, as well as apes and humans wherever they live. Although a typical mammal has two kinds of color detectors, most anthropoids have three: called *trichromatic* (three-color) vision. One form of trichromatic vision developed in the ancestor of all Old World primates, and a different form evolved convergently in several kinds of New World monkeys. In both groups, the color detectors are concentrated in the fovea, which enables the perception of powerfully sharp, color images. So, when *The Wizard of Oz* changes from the sepia monochrome of Kansas to the technicolor world of Munchkinland, the movie repeats something like what happened during the evolution of anthropoid vision.*

Big bodies, big brains

Primates, as a group, have large brains relative to their body weight, and this trait became accentuated in anthropoids, especially as they became larger animals. A recent evolutionary analysis[3] showed that brain expansion didn't occur just once in anthropoids, as previously thought. Instead, larger brains not only evolved in ancestral anthropoids, but also convergently in certain New World monkeys, in several kinds of Old World monkeys, in apes, and in humans.[3, 4]

Brain expansion is only part of the story, though. As we explained in Chapter 2,† several new cortical areas developed in anthropoids as their brains expanded, including new parts of the temporal, parietal, and frontal lobes. According to one estimate,[5] anthropoids have approximately 130 cortical areas; the last common

---

* *The Wizard of Oz* used a technicolor process that recorded three channels of light—on three strips of black-and-white film. By using light splitters and color filters, each strip captured the intensity of a different color of light, and a composite print of these negatives resulted in full-color cinematography.
† In the sections entitled "Primates go out on a limb" and "Anthropoids arrive."

ancestor of primates and rodents probably had fewer than 30. The new areas that emerged in anthropoids provided them with several additional kinds of representations, each of which gave them an advantage over their ancestors. In the next chapter, we'll deal with the new frontal areas of anthropoids; here we'll focus on the temporal and parietal areas—mainly the former.

New cortical areas augmented the older ones that anthropoids inherited from earlier ancestors. In the next section, we'll highlight two of these older areas: the perirhinal cortex and the hippocampus. The perirhinal cortex originated in early mammals; the hippocampus arose much earlier, in ancestral vertebrates.* Our emphasis on older cortical areas might seem strange for a chapter on the evolutionary innovations of anthropoids. We'll get to the new areas eventually, but to understand them it'll be useful to discuss some older areas first.

The qualities of vision

Expert opinion about the functions of the hippocampus and the perirhinal cortex has undergone a radical revision in recent years. It was once believed that both areas function exclusively in memory and that other brain areas underlie perception. This idea made sense at the time because some people with amnesia—who have damage to their hippocampus and perirhinal cortex—seem to perceive the world normally despite their memory loss. This impression proved to be mistaken, however. Recent research has revealed that both the hippocampus and the perirhinal cortex function in perception as well as memory; and they perform these functions differently because they specialize in representing distinctive kinds of feature conjunctions.

Bugs and blobs for brains

In a key experiment,[6] people saw pairs of either bugs, beasts, barcodes, or blobs, and in Figure 7.2a we illustrate two of each. The experimenters randomly designated one member of each pair as correct and the other as incorrect. In Figure 7.2c, we show which is which for some of the bugs. Sometimes, the two items had many features in common, in which case they had a high level of feature overlap, thereby creating high feature ambiguity. At other times, the two items had fewer features in common, in which case they had low or intermediate feature ambiguity.

Take bugs for example. In Figure 7.2c, we illustrate a series of four trials, arranged from top to bottom. Each trial required people to choose between a correct

---

* See the section of Chapter 2 entitled "The end of the brain."

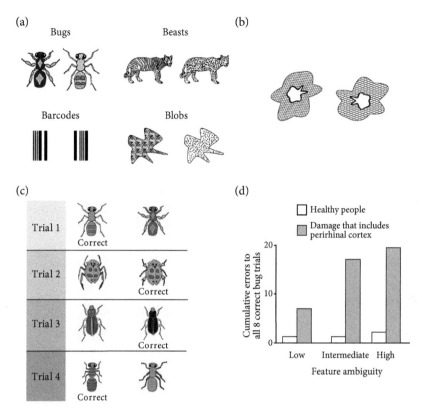

Fig. 7.2 The bugs, beasts, barcodes, and blobs task. (a) A pair of each type of item, as used in a memory test. (b) A different kind of blob, from another experiment. (c) A series of four trials for the bug test. People had to learn which member of each pair was designated as correct. Here, we mark the correct choice, but in the experiment the participants had to learn the correct choice by trial-and-error. (d) The white bars show scores for healthy people at each level of feature ambiguity. The gray bars show results from patients who had damage to both their hippocampus and their perirhinal cortex, as well as some nearby areas and connections.

Parts a, c, and d: Adapted from Morgan D. Barense, Timothy J. Bussey, Andy C. H. Lee, Timothy T. Rogers, R. Rhys Davies, Lisa M. Saksida, Elisabeth A. Murray and Kim S. Graham, Functional Specialization in the Human Medial Temporal Lobe, *The Journal of Neuroscience*, 25 (44), pp. 10239–10246, Figure 2, doi.org/10.1523/JNEUROSCI.2704-05.2005 ©The Society for Neuroscience.

Part b: Adapted from Morgan D. Barense, Iris I. A. Groen, Andy C. H. Lee, Lok-Kin Yeung, Sinead M. Brady, Mariella Gregori, Narinder Kapur, Timothy J. Bussey, Lisa M. Saksida, and Richard N. A. Henson, Intact Memory for Irrelevant Information Impairs Perception in Amnesia, *Neuron*, 75 (1), p. 157–67, doi: doi.org/10.1016/j.neuron.2012.05.014 © 2012 The Authors. This work is licensed under the Creative Commons Attribution License (CC BY). It is attributed to the authors.

bug and an incorrect one. At first, they could only guess which bug to choose, and they received immediate feedback if they made the correct choice. They would then perform another trial, with two new bugs. After the presentation of eight pairs of bugs, the original pair would reappear, and the experiment continued in this way until the people could identify the correct bug in each pair. High amounts of feature ambiguity made the test difficult, and the experimenters required people to pass a very stringent test: eight consecutive correct choices. Success required remembering a combination of visual features about body shape, marking patterns, and legs.

The rectangular bars in Figure 7.2d show the number of errors accumulated before passing the test. The gray bars come from patients with damage to their perirhinal cortex, hippocampus, and some nearby brain areas; the white bars come from healthy people. As the amount of feature ambiguity increased from low to high levels, the patients made more and more errors compared to healthy people, and the tests with beasts, barcodes, and blobs yielded similar results.[6] This sensitivity to the amount of feature overlap shows that the errors resulted from problems in perceiving (and distinguishing) items that share many features.

Even though the perirhinal cortex, the hippocampus, and other areas sustained damage in these patients, we know that their impairments resulted from a loss of perirhinal cortex function. As we'll explain in a while: (1) brain-imaging studies show significantly increased activation in the perirhinal cortex when people perceive and remember items with a high level of feature ambiguity; (2) selective removal or inactivation of the perirhinal cortex causes impairments in monkeys and rodents that resemble those in humans; and (3) similar disruption of other nearby areas, including the hippocampus, doesn't.

Another experiment in humans provided further support for these conclusions. Two blobs appeared simultaneously, and people tried to tell them apart.[7] As we illustrate in Figure 7.2b, each blob had three visual features—an outer shape, a different inner shape, and a fill pattern in between. Blobs could differ in one, two, or all three of these features. Those differing in just their fill pattern, for example, shared their inner and outer shapes and therefore had a high degree of feature ambiguity. In this situation, healthy people carefully scrutinized each blob for a long time, rather than looking rapidly from blob to blob as they would if they compared one feature at a time. This observation showed that the healthy people dealt with each three-feature blob as a whole. As they performed this task, a brain-imaging study revealed higher levels of activation in the perirhinal cortex when two blobs had a high level of feature ambiguity compared to blobs with a low level. What's more, patients with damage to their perirhinal cortex performed poorly on this test.[8] Again, sensitivity to the degree of feature overlap points to a perceptual impairment in these patients, mediated by the same cortical area that stores these feature conjunctions as memories.

Taken together, these results support the idea that the perirhinal cortex specializes in representations that distinguish visual items with a high level of feature ambiguity. The same results show that these representations contribute both to perception and memory: perception because of the need to discriminate different items from each other; memory because of the requirement for remembering which item had been correct in the past. The same conclusions apply to the hippocampus, but with an important twist. Instead of distinguishing objects and faces that have a high level of feature ambiguity, as representations in the perirhinal cortex do, representations in the hippocampus distinguish visual scenes.

This conclusion comes, in part, from experiments that used the *odd-one-out test*[9] (Fig. 7.3). People saw a set of four novel items on each trial, but three of the items came from the same source. The task rule required that they pick out the item that differed from the other three. Taking faces as an example, three of the items consisted of the same face viewed from different angles, along with the face of someone else. Unlike the tests we illustrated in Figure 7.2c, passing this test did not involve the memory of previously seen items. Therefore, it tested perception directly without calling on long-term memories (except for the task rule). The experimenters used three types of complex pictures in

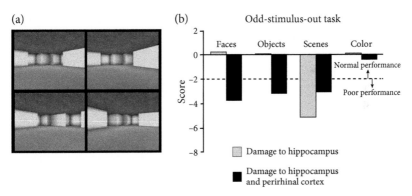

Fig. 7.3 The odd-one-out task. (a) Four visual scenes. This test required people to choose the scene—located in the lower left in this example—that differed from the other three. Otherwise identical tests used objects, faces, or colors. (b) Scores for two groups of patients: gray bars for patients with damage restricted to the hippocampus; black bars for patients with damage to both the hippocampus and the perirhinal cortex (along with some nearby areas). The more negative the score, the larger the impairment.

Adapted from Andy C. H. Lee, Morgan D. Barense, and Kim S. Graham, The Contribution of the Human Medial Temporal Lobe to Perception: Bridging the Gap between Animal and Human Studies, Quarterly Journal of Experimental Psychology, 58 (3–4b), pp. 300–325, Figure 2d, doi.org/10.1080/02724990444000168 Copyright © 2005, © SAGE Publications. Reprinted by Permission of SAGE Publications, Ltd.

this experiment: faces, objects, and scenes. In addition, they presented simple patches of color, which didn't require the representation of feature conjunctions. In Figure 7.3b, we display the results. Each bar shows the score of a group of patients with cortical damage. A score greater than –2 indicates a normal ability to pass the test; scores less than –2 indicate varying degrees of impairment. The biggest gray bar shows that when each visual item consisted of a large integrated visual scene, like the four scenes in Figure 7.2a, patients with damage to the hippocampus failed the test. The small gray bars indicate that these patients passed the test when the items consisted of colored squares, faces, or objects. Another group of patients had damage that extended from the hippocampus to include the perirhinal cortex, among other areas. The black bars reveal that these patients failed the test for faces, objects, and visual scenes, but passed for colored squares.

Figure 7.4 comes from a related experiment. In Figure 7.4a, we show examples of the kinds of stimuli used in the experiment: dot patterns, faces, and scenes. As illustrated in Figure 7.4b (beginning at the upper left), a face might be followed by a complex visual pattern and then by a second face. Participants tried to recognize whether the two faces differed.[8] In a process called *perceptual learning*, prior exposure to a given kind of stimulus improves the subsequent ability to distinguish such stimuli from each other. That exposure, which involved passive viewing of stimuli, occurred shortly before the test shown here. As shown in Figure 7.4c, a patient with damage to the hippocampus didn't benefit at all from prior exposure to visual scenes. The same patient did, however, benefit normally from prior exposure to dot patterns and faces. (Scores of healthy people are indicated by the gray shading labeled "normal range.") Figure 7.4d comes from a different patient, who had damage that involved both the hippocampus and the perirhinal cortex. In this patient, the impairment expanded to include faces as well as scenes.[8, 10]

Taken together, these findings show that damage to the perirhinal cortex or the hippocampus can either impair perception or not, depending on whether a certain test requires their respective representations. The hippocampus supports representations that distinguish visual scenes from each other, while the perirhinal cortex supports representations that distinguish faces from each other. The same kinds of representations also help people recognize a wide variety of complex objects.

Brain-imaging results support these conclusions. As we mentioned earlier, one experiment showed that the perirhinal cortex became highly activated when people viewed blobs that had a high degree of feature ambiguity, but not when blobs had less ambiguity. In another experiment, the hippocampus became significantly activated when people viewed scenes, but the perirhinal cortex did so when they saw faces.[8] Likewise, the pattern of hippocampal activations indicated what scene a person had viewed, but not what object,[11] and researchers have obtained

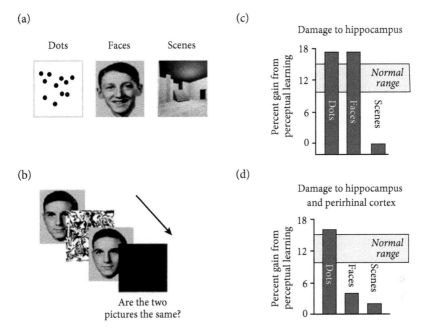

Fig. 7.4 Perceptual learning. The term perceptual learning refers to the influence of prior exposure to items on the subsequent memory of those items or items like them. (a) An example of each of three different kinds of visual items used in these experiments. (b) A sequence of pictures that participants saw—in the order indicated by the arrow. The first and third screens always displayed the same kind of item, faces in this example. The test required people to indicate whether the third item differed from the first. (c) Scores of a patient with damage restricted to the hippocampus. The height of each bar indicates the degree to which prior exposure to dots, faces, or scenes improved the ability to pass each test. The gray rectangle in the background indicates the range of scores obtained by healthy people. (d) Results for a patient with damage that included both the hippocampus and the perirhinal cortex, along with some nearby areas and connections.

Adapted from Matthew E. Mundy, Paul E. Downing, Dominic M. Dwyer, Robert C. Honey and Kim S. Graham, A Critical Role for the Hippocampus and Perirhinal Cortex in Perceptual Learning of Scenes and Faces: Complementary Findings from Amnesia and fMRI, *The Journal of Neuroscience*, 33 (25), pp. 10490–10502, Figure 2, doi.org/10.1523/JNEUROSCI.2958-12.2013 ©The Society for Neuroscience, 2013.

the opposite results for the perirhinal cortex.[12] Finally, the perirhinal cortex became highly activated when people distinguished animals with high feature ambiguity, such as sheep versus lambs, but not for animals with low ambiguity, such as pigs versus snakes.[13]

These findings all show that the hippocampus and perirhinal cortex specialize in different kinds of representations and that they function in both perception and memory. The same principle applies to other cortical areas. So far, we've concentrated on the perceptual functions of cortical areas previously thought to specialize in memory. The opposite also holds true: "sensory areas" also store memories.*

One relevant experiment had three phases.[14] First, people made judgments about whether they recognized famous people, particular locations, or how recently they had used an object. The experimenters could then use the pattern of activation in a variety of "sensory areas" to determine which kind of picture a person had perceived at any given time. Next, the same people learned pairings among the same pictures. The pictures in a pair didn't have anything to do with each other; the scientists simply chose two pictures arbitrarily and asked people to remember that they "went together." Finally, the same people made a judgment about whether two pictures—separated in time—"went together." During the interval between the two pictures, as people remembered the picture associated with the one they'd just seen, the pattern of brain activation from the first phase of the experiment rematerialized.

This finding shows that "sensory areas" of cortex have a similar pattern of activation when people recall the memory of a picture and when they perceive that picture in real time. Additional experiments show the same thing for objects, faces, scenes, and simple visual stimuli. For example, when people perceive how a line is tilted on a video screen, the same cortical areas became activated for the perception and memory of that tilt.[15] In one brain-imaging study, people saw and had to remember the orientation of lines in a visual grating. By studying the activation patterns in seven visual areas of cortex, scientists could reconstruct both the contents of memory and incoming visual inputs.[16] In another experiment, monkeys needed to detect familiar images, as opposed to novel ones. As usual, they tended to forget images that they hadn't seen for a long time. A group of neurons in their inferior temporal cortex responded less for familiar images, which could only happen if this area stored a representation of previously viewed images. The duration of this memory signal, called *repetition suppression*, closely matched the ability of the monkeys to detect an image's familiarity.[17] In yet another study, scientists recorded neuronal activity from 42 cortical areas in monkeys. They concluded that most of the cortical areas specializing in visual representations contribute to both online visual processing and memory.[18]

Taken together, these and other experiments show that representations in "sensory areas" support both memory and perception, just as representations in the hippocampus and perirhinal cortex do. When you recollect a memory, you probably do so by reinstating the patterns of neuronal activity that occurred when you

---

* More than two dozen visual areas evolved in anthropoids, and many additional areas represent sounds or contact with the body. Collectively, scientists usually call them sensory areas because they underlie perception, but because they also store memories, we place "sensory areas" in quotation marks throughout this chapter.

experienced something for the first time. In the next two sections, we summarize some experiments on monkeys and rodents that lead to the same conclusions.

Morphing for monkeys

Experiments on monkeys have an important advantage over human studies: scientists can remove or inactivate specific brain areas. For example, we've said that the perirhinal cortex represents faces and objects because damage to this cortical area in people causes a specific pattern of perceptual and mnemonic impairments. Unfortunately, the brain damage in these patients often extends to nearby cortical areas and may disrupt connections that might have nothing to do with the perirhinal cortex. Indeed, when people suffer brain damage it almost always involves several brain regions. So, it's helpful to corroborate our conclusions in animal experiments. The selective removal or inactivation of the perirhinal cortex in monkeys has provided this validation.[19] What's more, the interplay between human and animal experiments runs both ways. Some of the most influential experiments in the previous section were, in fact, inspired by tasks first used to study perception and memory in monkeys. Without those experiments, scientists might never have discovered the role of both the perirhinal cortex and hippocampus in perception.

One study of the perirhinal cortex in monkeys produced results like those illustrated in Figure 7.2d for humans,[19] but another one provided particularly decisive results.[20] After monkeys had learned which of two pictures to choose, scientists generated morphed intermediates between the two images. The monkeys had easily distinguished between the original pictures because they had very little feature ambiguity. The morphed images, in contrast, had either high or low amounts of feature ambiguity, and the monkeys hadn't seen them prior to testing. After a surgical removal of their perirhinal cortex, monkeys failed to distinguish images with high—but not low—feature ambiguity.

Scientists have also used the odd-one-out test on monkeys.[21] The monkeys saw pictures of one object from five different viewing perspectives, along with the picture of a different object: the odd-one-out. To get some food, they had to choose the odd-one-out, regardless of where it appeared in the array of six images. Like the results from humans, monkeys needed an intact perirhinal cortex to pass this test with objects and faces, but they performed as well as normal monkeys on difficult oddity judgments for color, size, and geometric shapes.

Another monkey experiment contrasted the function of the perirhinal cortex with that of the hippocampus.[22] In Chapter 1,* we explained that memory scientists

---

* In the section entitled "The second road."

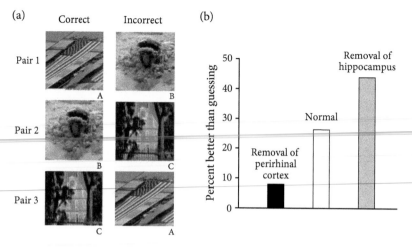

Fig. 7.5 A monkey memory test. (a) Three pairs of pictures. For each pair, there is one correct choice, which appears in the left column. (b) The effect of removing either the perirhinal cortex (black bar) or the hippocampus (gray bar) compared to normal monkeys (white bar).

Adapted from Elisabeth A. Murray, Timothy J. Bussey, and Lisa M. Saksida, Visual Perception and Memory: A New View of Medial Temporal Lobe Function in Primates and Rodents, *Annual Review of Neuroscience*, 30, pp. 99–122, Figure 6, doi.org/10.1146/annurev.neuro.29.051605.113046 © Annual Reviews, 2007.

once believed that these two areas work together as parts of a single "medial temporal lobe memory system." The results illustrated in Figure 7.5 demonstrated, to the contrary, that the perirhinal cortex and hippocampus have different functions. The task involved three pictures, labeled A, B, and C in Figure 7.5a. The monkeys saw two of these pictures at a time. Given the choice between pictures A and B, the monkeys needed to choose A to get food; and given the choice between B and C, they needed to choose B. These two pairings seemed to establish a simple sequence: picture A should have been the correct choice over C. But it wasn't. Given the choice between pictures A and C, the monkeys needed to choose C. Accordingly, all three pictures had the same value because they all served as correct and incorrect choices in equal measure. The white bar in Figure 7.5b shows the scores obtained by normal monkeys. The black bar shows that, after a surgical removal of their perirhinal cortex, monkeys barely exceeded (by just 9%) what they could get by guessing. The gray bar shows that after removal of the hippocampus, monkeys performed significantly better than normal, reaching 95% correct.*

---

* You might wonder how brain damage could improve memory. The hippocampus represents visual scenes, so it helps monkeys appreciate the arrangement of items in a scene. In this task, the arrangements are irrelevant, so removing the hippocampus decreases any distraction caused by this information.

Because removing one area makes monkeys worse and removing the other makes them better, we can be sure that the perirhinal cortex and the hippocampus perform different functions.

One last monkey experiment focused on the inferior temporal cortex, a highly developed part of the visual cortex in anthropoids. These areas lie between the occipital lobe and the perirhinal cortex. Removal of the inferior temporal cortex caused an impairment in distinguishing colors but not objects; removal of the perirhinal cortex produced the opposite results.[23]

In addition, studies of brain activation or neuronal activity too numerous to mention reveal two additional similarities between human and monkey brains: inferior temporal areas represent color and shape conjunctions of intermediate complexity; and the perirhinal cortex and the hippocampus represent conjunctions of high complexity, such as unique objects and scenes.

Rodents wrongly recognizing

Rodents lack most of the inferior temporal areas of anthropoids, but both groups of species have a perirhinal cortex and a hippocampus. The specialization of the hippocampus for scene representations, as discussed earlier in this chapter, has an obvious relationship to the navigational functions that we considered in Chapter 4 for the rodent hippocampus. The function of the perirhinal cortex also seems to be similar in rodents and primates. For example, its removal caused an impairment in the ability of rats to make odd-one-out judgments, but only when the items shared many features,[24,25] just like the results from humans and monkeys. In another experiment, removal of the perirhinal cortex in mice eliminated the unique representations of objects, as revealed by their inability to tell whether an object was familiar or novel.[26] In fact, their memory of objects was so bad that they sometimes treated novel objects as if they were familiar, an impairment called *false recognition*. We could say a lot more about the perirhinal cortex of rodents, but these two examples suffice to show that its representations resemble those in humans and monkeys.

Anthropoid accounting

To this point, we've emphasized perceptions and memories of the *qualitative* aspects of vision, features such as color, shape, and surface appearance. Cortical areas in the temporal lobe specialize in these representations. Cortical areas in the parietal lobe represent the *quantitative* aspects of vision: number, size, location (along with other spatial information), speed, distance, duration, and order, among others.

For example, monkeys have learned to choose a display with either a larger number of items or a smaller number, depending on a task rule.[27] In other studies, monkeys made choices based on the distance to an item.[28] Dozens of experiments on anthropoids have revealed brain activations and patterns of neuronal activity indicating that the parietal cortex represents quantities.[29]

There's much more to say about metrics and the parietal cortex, but it's time to sum up this chapter.

Anthropoid additions reconsidered

The idea that each cortical area specializes in a particular kind of representation has some revolutionary implications. For more than a century, scientists have categorized cortical areas using functional terms such as perception, memory, the control of movement, or executive control, concepts that have little to do with each other. In contrast, our idea puts the entire cerebral cortex on an even footing. Cortical areas differ in terms of their representations, which natural selection can modify in the same way as it does for other traits, such as the number and structure of toes or teeth.

Our idea also puts large-scale neural circuits in a new perspective. Often called *connectomes*—to bask in the reflected glory of the genome—large-scale circuits form a network of neurons spanning many cortical areas. As information flows through a large-scale cortical network, smaller networks in each area pick off a different combination of features, which establishes representations of a particular type at a particular level of complexity. The anthropoid cortex has a well-studied example. It's possible to trace the flow of visual information from the occipital lobe through inferior temporal areas to the perirhinal cortex. To many scientists, this sequence seems to be aimed at the representation of objects; they consider the feature conjunctions represented in the intervening, inferior temporal areas to be simply a means to that end. Taking evolution into account, however, we can appreciate that each level of representation—from the most elemental to the most complex—contributed something to the evolutionary success of anthropoids.

It's all well and good to allude to vague evolutionary advantages, but what are they, specifically? We think that each new kind of representation provided a specific advantage for foraging. As anthropoids evolved from small animals into larger ones, they needed to travel long distances to obtain nutrients, often from fruiting trees, and they did so in daylight. The fovea that they inherited from an earlier primate ancestor, along with the trichromatic vision that evolved later, improved their ability to see, distinguish, and recognize combinations of colors and shapes at a distance. New cortical areas, mainly in the inferior temporal lobe, represented these distant sights, and the superior temporal cortex did something similar for distant sounds. The sights served as signs that a particular stand of trees might have ripe fruit, and so did the sounds made by birds and primates as they fed there. Largely in parallel, parietal areas represented quantitative features of the visual world, such as the number of food items at various places and their relative distances, how long it should take to get there, and so forth. All of these new representations contributed to the success of our anthropoid ancestors in long-distance, visually guided foraging.

These cortical innovations had a surprising consequence. In Chapter 6, we discussed some of the visual innovations of early primates. They used vision to guide movements with amazing accuracy, and these memories needed regular updating to match the physical realities of their arboreal world. Any inaccuracy in visual processing harmed their chances of survival. In anthropoids, vision evolved to tolerate inaccuracies because foraging choices, especially at a distance, often depended on relative quantities and qualities rather than absolute accuracy. Several visual illusions result from tricks that relative size or some other relative feature plays on perception. So, it's fair to say that such illusions result, in part, from the foraging life of ancestral anthropoids.

The new representations of anthropoids also served other functions, especially social ones. Early primates lived solitary lives for the most part and foraged at night.[3] All modern primates have descended from these solitary, nocturnal ancestors, so when large social groups evolved among diurnal anthropoids they had to "invent" a social system from scratch—or close to it. (Other mammals also developed large and complex social groups, but they did so convergently.) As diurnal animals, anthropoids could see each other quite well while foraging, so cortical representations that evolved for the detection and assessment of nutritional resources could also support the recognition of group members and social signals. For example, skin color is important for mate selection because it provides information about the health of an individual.[30] Notwithstanding the crucial role of vision in the social life of anthropoids, it seems likely that the need to feed drove the evolution of new visual representations in their neocortex. Why? Because neither of the main enhancements in primate vision occurred in concert with an increase in anthropoid group size: the primate fovea evolved long beforehand (in haplorhines); trichromatic vision evolved long afterward (in the founding population of Old World primates).

As they skip along the yellow brick road, Dorothy and her companions meet some of the most famous anthropoids in cinematic history. Neither *King Kong* nor the apes on the *Planet of the Apes* can do what monkeys do in *The Wizard of Oz*, which is fly. Because of their long, daytime journeys, our anthropoid ancestors ran some serious risks. Snakes, predatory birds, and large carnivorous mammals could ambush them at any time. These predators would have posed less danger if real anthropoids could fly like the winged monkeys in the Land of Oz. They couldn't do that, of course, but our anthropoid ancestors mitigated their risks in a different way—as we'll see in the next chapter.

References

1. Busigny, T., Joubert, S., Felician, O., Ceccaldi, M., & Rossion, B. Holistic perception of the individual face is specific and necessary: evidence from an extensive case study of acquired prosopagnosia. *Neuropsychologia* 48, 4057–4092 (2010).

2. Ross, C. F. in *Anthropoid Origins: New Visions* (eds C. F. Ross & R. F. Kay), pp. 477–537 (New York, NY: Academic/Plenum, 2004).

3. DeCasien, A. R., Williams, S. A., & Higham, J. P. Primate brain size is predicted by diet but not sociality. *Nature Ecology and Evolution* 1, 0112 (2017).

4. Gonzales, L. A., Benefit, B. R., McCrossin, M. L., & Spoor, F. Cerebral complexity preceded enlarged brain size and reduced olfactory bulbs in Old World monkeys. *Nature Communications* 6, 7580 (2015).

5. Kaas, J. H. The evolution of brains from early mammals to humans. *Wiley Interdisciplinary Reviews in Cognitive Science* 4, 33–45 (2013).

6. Barense, M. D. et al. Functional specialization in the human medial temporal lobe. *Journal of Neuroscience* 25, 10239–10246 (2005).

7. Barense, M. D. et al. Intact memory for irrelevant information impairs perception in amnesia. *Neuron* 75, 157–167 (2012).

8. Mundy, M. E., Downing, P. E., Dwyer, D. M., Honey, R. C., & Graham, K. S. A critical role for the hippocampus and perirhinal cortex in perceptual learning of scenes and faces: complementary findings from amnesia and fMRI. *Journal of Neuroscience* 33, 10490–10502 (2013).

9. Lee, A. C., Barense, M. D., & Graham, K. S. The contribution of the human medial temporal lobe to perception: bridging the gap between animal and human studies. *Quarterly Journal of Experimental Psychology B* 58, 300–325 (2005).

10. Barense, M. D., Gaffan, D., & Graham, K. S. The human medial temporal lobe processes online representations of complex objects. *Neuropsychologia* 45, 2963–2974 (2007).

11. Bonnici, H. M. et al. Decoding representations of scenes in the medial temporal lobes. *Hippocampus* 22, 1143–1153 (2012).

12. Libby, L. A., Hannula, D. E., & Ranganath, C. Medial temporal lobe coding of item and spatial information during relational binding in working memory. *Journal of Neuroscience* 34, 14233–14242 (2014).

13. Clarke, A. & Tyler, L. K. Object-specific semantic coding in human perirhinal cortex. *Journal of Neuroscience* 34, 4766–4775 (2014).

14. Lewis-Peacock, J. A. & Postle, B. R. Temporary activation of long-term memory supports working memory. *Journal of Neuroscience* 28, 8765–8771 (2008).

15. Harrison, S. A. & Tong, F. Decoding reveals the contents of visual working memory in early visual areas. *Nature* 458, 632–635 (2009).

16. Rademaker, R. L., Chunharas, C., & Serences, J. T. Simultaneous representation of sensory and mnemonic information in human visual cortex. *BioRxiv*, doi: 10.1101/339200 (2018).

17. Meyer, T. & Rust, N. Single-exposure visual memory judgments are reflected in inferotemporal cortex. *Elife* 7, doi:10.7554/32259 (2018).

18. Dotson, N. M., Hoffman, S. J., Goodell, B., & Gray, C. M. Feature-based visual short-term memory is widely distributed and hierarchically organized. *Neuron* 99, 215–226 (2018).

19. Murray, E. A., Bussey, T. J., & Saksida, L. M. Visual perception and memory: a new view of medial temporal lobe function in primates and rodents. *Annual Reviews in Neuroscience* 30, 99–122 (2007).

20. Bussey, T. J., Saksida, L. M., & Murray, E. A. Impairments in visual discrimination after perirhinal cortex lesions: testing 'declarative' vs. 'perceptual-mnemonic' views of perirhinal cortex function. *European Journal of Neuroscience* 17, 649–660 (2003).

21. Buckley, M. J., Booth, M. C. A., Rolls, E. T., & Gaffan, D. Selective perceptual impairments after perirhinal cortex ablation. *Journal of Neuroscience* 21, 9824–9836 (2001).

22. Saksida, L. M., Bussey, T. J., Buckmaster, C. A., & Murray, E. A. Impairment and facilitation of transverse patterning after lesions of the perirhinal cortex and hippocampus, respectively. *Cerebral Cortex* 17, 108–115 (2007).

23. Buckley, M. J., Gaffan, D., & Murray, E. A. Functional double dissociation between two inferior temporal cortical areas: perirhinal cortex versus middle temporal gyrus. *Journal of Neurophysiology* 77, 587–598 (1997).

24. Bartko, S. J., Winters, B. D., Cowell, R. A., Saksida, L. M., & Bussey, T. J. Perceptual functions of perirhinal cortex in rats: zero-delay object recognition and simultaneous oddity discriminations. *Journal of Neuroscience* 27, 2548–2559 (2007).

25. Bartko, S. J., Winters, B. D., Cowell, R. A., Saksida, L. M., & Bussey, T. J. Perirhinal cortex resolves feature ambiguity in configural object recognition and perceptual oddity tasks. *Learning and Memory* 14, 821–832 (2007).

26. McTighe, S. M., Cowell, R. A., Winters, B. D., Bussey, T. J., & Saksida, L. M. Paradoxical false memory for objects after brain damage. *Science* 330, 1408–1410 (2010).

27. Brannon, E. M. The representation of numerical magnitude. *Current Opinion in Neurobiology* 16, 222–229 (2006).

28. MacDonald, S., Spetch, M. L., Kelly, D. M., & Cheng, K. Strategies in landmark use by children, adults and marmoset monkeys. *Learning and Motivation* 35, 322–347 (2004).

29. Genovesio, A., Wise, S. P., & Passingham, R. E. Prefrontal-parietal function: from foraging to foresight. *Trends in Cognitive Sciences* 18, 72–81 (2014).

30. Henderson, A. J. et al. Skin colour changes during experimentally-induced sickness. *Brain Behavior and Immunity* 60, 312–318 (2017).

8

Anthropoid augmentations

To eat or be eaten

DOROTHY: I don't like this forest . . . it's dark and creepy!
SCARECROW: Do you suppose we'll meet any wild animals!
TIN MAN: We might . . .
SCARECROW: Animals that eat straw?
TIN MAN: Some, but mostly lions, and tigers, and bears!

Oh my! A bear might eat a backpacker every now and then, and sharks sometimes snack on the odd surfer, but people rarely get eaten nowadays. In the novel *Jurassic Park*, the author imagines what it feels like to become another animal's food. After being incapacitated by a dinosaur that squirts a burning toxin into his eyes, the victim felt:

> new pain on both sides of his head. The pain grew worse, and as he was lifted to his feet he knew the dinosaur had his head in its jaws, and the horror of that realization was followed by a final wish, that it would all be ended soon.
>
> —Michael Crichton, *Jurassic Park*
> Excerpt from JURASSIC PARK: A NOVEL by Michael Crichton,
> copyright © 1990 by Michael Crichton. Used by permission of Alfred A. Knopf,
> an imprint of the Knopf Doubleday Publishing Group, a division of Penguin Random
> House LLC. All rights reserved.

Unlike modern humans, our anthropoid ancestors risked predation regularly. No one knows how frequently they succumbed, but two experts on vervets—a typical anthropoid monkey—estimated that more than half of them get eaten before they reach adulthood.[1] Regardless of the exact predation rate, life in the wild posed grave dangers for our anthropoid ancestors. The first anthropoids* were small animals, and, unlike early primates, they foraged in daylight. Even if they did so at dusk and dawn, they faced more peril than did feeders of the night. Despite these dangers, anthropoids survived and thrived.

In this chapter, we'll discuss some evolutionary innovations that mitigated their risks. New parts of the prefrontal cortex turbocharged their ability to learn quickly

---

* In Figure 2.4, we illustrate the relationship between anthropoids (green lines) and other primates (blue lines).

The Evolutionary Road to Human Memory. Elisabeth A. Murray, Steven P. Wise, Mary K. L. Baldwin, and Kim S. Graham, Oxford University Press (2020). © Oxford University Press.
DOI: 10.1093/oso/9780198828051.001.0001

and cope with novel situations. So, they made fewer mistakes when setting out on long, dangerous, daytime journeys for food.[2] By reducing such errors, anthropoids could obtain enough nourishment while limiting their exposure to predators.

Global cooling

About 34 million years ago—long after anthropoids emerged and long before the appearance of the first apes—life became more difficult for anthropoids. At about that time, the Earth cooled. On a global scale, temperatures fell by approximately 2°C (3.6°F) in a couple of million years, which is a lot for that sort of thing. This amount of climate change probably disrupted food supplies in the tropics, where nearly all anthropoids lived at the time.

Evidence from tooth-wear patterns suggests that these animals preferred a diet of fruit, which has a high nutritional value.[3] Unfortunately for our ancestors, fruiting patterns are especially sensitive to climate change. Even in a stable climate, fruiting is highly variable among tree species and among individual trees of any given species.[4] Global cooling exacerbated these problems, so fruit-eating anthropoids became susceptible to famine. They had the option of turning to fallback foods, such as leaves, tree sap, and insects, which remained prevalent and provided essential vitamins and amino acids. But these foods couldn't provide enough energy for the large, active animals that most anthropoids had become.*

After the New World and Old World branches of the anthropoid family tree separated, the brain expanded in both lineages as new cortical areas emerged. In Chapter 2,† we explained that anthropoid evolution added several prefrontal areas, and, in this chapter, we'll explore how they contributed to the success of anthropoids.

Knowing but blundering

In the early 1970s, Hans-Lukas Teuber[5] proposed that the prefrontal cortex uses the memory of previous errors to reduce future ones. He noted that:

> Patients with frontal lobe disease . . . seem to perceive the mistakes they make, but are unable to use the information to guide their behavior.

In one case, Teuber reported what happened when patients with damage to their prefrontal cortex performed the Wisconsin card sorting task. These patients saw

---

* In *Baboon Metaphysics: The Evolution of a Social Mind*, (University of Chicago Press, Chicago, IL, 2008), Dorothy L. Cheney and Robert M. Seyfarth survey the life of anthropoids.
† In the section entitled "Anthropoids arrive."

cards that had a variable number of colored shapes. For example, one card might have four red circles; the next might have one green star. The patients then placed each card—one at a time—into piles sorted by either color, number, or shape. For the card with a single green star, they had to decide whether it belonged in the pile of cards that each had a single shape or in a pile of cards with green shapes. They didn't know the sorting rule at first; they simply had to guess and wait for an examiner to tell them if they'd gotten it right. Teuber offered the following anecdote:

> There is the patient who . . . proceeds to say (correctly): "You are probably starting with color—so that if I put this card down, it will be wrong? You see—I am right—this one is wrong! And this one—wrong! and wrong again!" He then proceeds in this fashion by contradicting in his actions what he can announce verbally as the correct procedure, evidently aware of the contradiction but incapable of avoiding it.[6]

The idea that the prefrontal cortex functions to reduce errors incorporates many other ideas about these cortical areas, such as a role in representing "goals and the means to achieve them"[7]: achieve a goal, prevent an error. Likewise, many experts say that the prefrontal cortex has "executive functions," a concept that refers to deciding and planning what to do.

According to one recent idea,[2] the anthropoid prefrontal cortex reduces errors because its neurons combine and store information related to goals, in much the same way as neurons in visual areas of cortex combine and store information about shapes and colors. The perirhinal cortex, for example, represents unique conjunctions of visual features that characterize objects and distinguish them from each other, as we discussed in Chapter 7.* The anthropoid prefrontal cortex does something similar, but in addition to visual features its conjunctions incorporate "what to do" in a given context, including a goal and an action that achieves that goal, along with an outcome that's likely to follow. A goal, in this sense, corresponds to the target of an action, such as an object to obtain; the term outcome refers to predictable follow-on events. These complex, multicomponent representations— combinations of contexts, goals, actions, and outcomes—correspond to memories of what we call *discrete goal-related events*.

For example, imagine that you're at home (a context) when your smartphone emits a ringtone, which in this case identifies a call from your best friend. This contextual information elicits a memory that includes the goal of grasping the phone (an action) and swiping the touchscreen (another action) in the expectation of hearing your best friend's voice (an outcome). Somewhere in your prefrontal cortex, a network of neurons represents the combination of all of these elements: collectively, a discrete goal-related event. This unique representation contrasts

---

* In the section entitled "The qualities of vision."

with other, similar memories. If you're in a theater when you hear the ringtone, the associated action would involve silencing the phone rather than answering it.

Neuroscientists have accumulated evidence about the feature conjunctions that the prefrontal cortex represents. Some of its neurons represent a specific combination of sensory features linked to a given action.[8] Others represent the combination of a particular sensory context and either a specific problem-solving strategy or a spatial goal.[9] Additional representations include the conjunction of a particular action and an outcome,[10-12] including its desirability, probability, and magnitude.[13,14] A particularly good example is the conjunction between the sensory features of an object and whether it has been chosen previously during a series of trials. This conjunction can be held in memory to guide future choices until the series ends.[15] And this very brief summary mentions only a few of the feature conjunctions represented in the anthropoid prefrontal cortex.

Taken together, representations in the anthropoid prefrontal cortex augment those in the "sensory areas" of cortex by adding various combinations of goal-, action-, and outcome features to the features that define a given sensory context. When that context recurs, anthropoids can use these representations to recall a goal, an action, and an outcome—all based on the memory of a discrete goal-related event. Through connections with premotor areas, these representations generate a bias toward the represented action, as we discussed in Chapter 6.* For example, neurons representing the conjunction of an object's sensory features and an especially beneficial outcome can enhance the neural circuits that guide a reaching movement to that object (at the expense of circuits that would guide reaching to a different object).†

Pay by paragraph

Given the high predation risks and resource volatility faced by ancestral anthropoids, any shift from fruitless foraging excursions to successful ones surely contributed to their survival.[2] When resources became scarce and volatile, especially during the dramatic global cooling of 34 million years ago, the older forms of memory could not adapt quickly enough. The reason is that the kind of memories we discussed in Chapter 3—Pavlovian and instrumental associations—are cumulative forms of memory that store average outcomes over several events, rather than particular outcomes that occur during discrete events.

---

* In the section entitled "Deciding on desirability."

† This way of making choices avoids the requirement for a little "decider" inside the brain, sometimes called a homunculus. The representation of an action associated with the best behavioral outcome will win a competition and thereby control behavior, with no need for a homunculus—at least for some decisions.

To understand why averages adapt slowly, imagine that you're being paid to proof-read these paragraphs. At first, you receive $1 per paragraph, but after a while the pay rate jumps immediately to $8. If, after the first $8 paragraph, someone asks you how much the previous paragraph paid, your answer should be $8, of course. This answer requires the memory of a discrete event: proof-reading a paragraph and pocketing $8. But what if you can't remember the $8 payment and instead incorporate it into an average? If the average spans seven events, then after one $8 paragraph the average payment only changes from $1 to $2—$6 short of the current pay rate. Even if the average gives more weight to the most recent payments, it still takes seven paragraphs to reach the right answer ($8 per paragraph).

To many scientists, the ability to remember discrete goal-related events doesn't seem like such a big deal. As we explained in Chapter 3, all animals learn associations among sensations, actions, and outcomes, and these memories all depend on prior events. Surely, the emergence of new prefrontal areas should lead to something more dramatic: a new form of reasoning, perhaps. But evolution can work in more subtle ways. By augmenting older, average-based forms of memory with novel representations of discrete goal-related events, the new prefrontal areas of our anthropoid ancestors helped them limit the number of bad foraging choices. In the pay-by-paragraph example, after the pay rate changes from $1 to $8, your memory of a recent $8 payment (a discrete event) can be used along with a memory of the average payment ($2 at first). The event memory provides more motivation to read the next paragraph. The same thing happens in reverse, of course, when values diminish. In their natural habitat, the new prefrontal areas of anthropoids helped them remember coming up empty after a long journey to a previously productive cluster of trees. An ancestral anthropoid could use the memory of this event to avoid another risky and (literally) fruitless excursion to the same place. Because bad foraging choices risk predation (not to mention the wasted energy), "eat but don't get eaten" is a maxim that maximizes evolutionary success.

Monkey business

Although we can't study the anthropoids of 34 million years ago, we can examine memory in modern species. Nearly all of this research involves rhesus monkeys, which make such a nuisance of themselves in their native India that many villages and cities employ monkey catchers to mitigate the damage. Rural monkeys destroy crops worth hundreds of millions of dollars, and urban monkeys run wild, biting a thousand people every day:

> Wherever they go, panic spreads . . . Any houses . . . raided by monkeys (are left) in shambles—eatables on the floor, crockery broken, taps open, wires cut, plants mauled.[16]

Knowing rhesus monkeys as we do, this description of their destructiveness seems charitable. In 2007, a Delhi politician fell to his death attempting to escape a hoard of rhesus monkeys that attacked him on an elevated terrace.[17] All of this goes to show that rhesus monkeys forage fiercely and effectively, even in situations far afield from the habitats in which they evolved.

Laboratory experiments show that the prefrontal cortex of rhesus monkeys speeds learning and reduces errors. To illustrate this point, we'll summarize the results of five experiments, which highlight just two parts of the prefrontal cortex, both of which evolved in anthropoids: the *ventral prefrontal cortex* and the *dorsolateral prefrontal cortex*.* The first one deals with a concept called *credit assignment*.

Remembering events

The concept of credit assignment refers to the link between a particular outcome, such as food availability, and the specific choice that led to it. The idea is that when a choice causes an outcome, it deserves to be remembered for doing so: credit where credit is due. For outcomes that follow a choice promptly and consistently, allocating credit is easy. But when a choice produces an outcome inconsistently or after a long delay, accurate credit assignment can be challenging.

A guessing game explains why. Let's say that you can choose door number 1, which produces a prize half the time, or door number 2, which does so 10% of the time. At first, you know nothing about these likelihoods. Imagine that you choose door 1 first, but no pay-off occurs, which is a 50–50 chance that doesn't come through. Based on this disappointment, you choose door 2 next. Amazingly, its 1-in-10 chance pays off. What should you do next? Door 2 seems best even though door 1 is five times more likely to yield a prize. It takes many choices to accurately assign door 1 the prize-producing credit that it deserves. Now imagine how this difficulty escalates if the pay-off percentages change during the game.

Cumulative forms of memory, such as Pavlovian and instrumental associations, have trouble keeping up to date with rapidly changing probabilities. They use the feedback from events to adjust their averages, but they don't store each discrete event. So, feedback changes the state of an existing memory, but it doesn't record a goal-related event in detail. This approach works well when outcome probabilities remain constant or change slowly. But as we explained in our pay-by-paragraph scenario, averages fall well behind the reality when pay-off percentages change rapidly.

---

* The anthropoid prefrontal cortex consists of several major subdivisions, each named according to its location in the frontal lobe. The term ventral (from Latin for "toward the belly") refers to something near the bottom of the brain. The term dorsolateral means that this area is near the top of the brain (dorsal) when viewed from the side and near the side of the brain (lateral) when viewed from above.

In a key breakthrough, a group of researchers led by Matthew Rushworth found that the prefrontal cortex helps monkeys assign credit to choices. It does so quickly and accurately, thus improving on average-based memories.[18] But what cortical area does this job, and when did it evolve? The first studies attributed this function to the orbital cortex,[18] which evolved in early primates, if not a little earlier. More recent research, however, has revealed that accurate credit assignment depends instead on the *ventral prefrontal cortex*,[19] which emerged during anthropoid evolution. We now know that these two cortical areas perform different and complementary functions. Both areas contribute to updating the valuation of predicted outcomes, but of different kinds. One depends on *desirability* valuations; the other involves *availability* valuations.

Two examples illustrate this distinction. No one involved in the dating scene needs to be reminded of the difference between desirability and availability. Without this distinction, you wouldn't know what settling means, but you do. Another example involves a choice among restaurants. Cost aside, a good restaurant is more desirable than a mediocre one. But if the likelihood of getting a table in the former is low, the latter might be the better choice, especially if you're really hungry.

Before discussing the credit-assignment experiment, recall a few key points from Chapter 6.* In Figure 6.4a, we show the hand of a primate in the act of choosing among three berries—all within reach. The activity of neurons in the temporal lobe provides information about the color and glossiness of each berry to the orbital cortex, which in turn assesses each berry's current desirability via interactions with the amygdala. These valuations depend, in part, on an animal's current state, such as being hungry for berries or having a belly full of them. Via its connections, neural interactions between the orbital cortex and the amygdala combine representations of an object's sensory features with the sensory features of an associated food and the value of that food to an animal in its current state (full of berries, for example). In Figure 6.5, we illustrate the *devaluation task* and what happens as an animal consumes berries to satiety. After eating enough berries, the prospect that a given choice will yield yet another berry becomes dreadful. So, monkeys usually make choices that yield some other kind of food. Results from the devaluation task reveal the fundamental function of the primate orbital cortex; it improves foraging choices by updating the *desirability valuation* of predicted outcomes.

The ventral prefrontal cortex does something different, something related to a food item's *availability* rather than its *desirability*. As we explained in Chapter 2,† several new temporal areas emerged during anthropoid evolution, many of which represent the visual signs of resources and their value. A stand of fruitful trees,

---

* In the section entitled "Deciding on desirability."
† In the section entitled "Anthropoids arrive."

for example, has a characteristic appearance when viewed at a distance, and the combination of its visual features serves as a sign of food. Likewise, at close range, a berry's glossiness signals something about its sweetness. In both instances, the temporal lobe relays this information to the ventral prefrontal cortex.

An experiment on monkeys revealed what this anthropoid area does with visual information of this kind. It used the *three-arm bandit task*, a name that alludes to the gambling machines called one-arm bandits (also known as slot machines).

On each trial, monkeys saw the same three pictures—A, B, and C—which exchanged locations from trial to trial. In Figure 8.1a, we illustrate the first four of 300 trials. Every time these three pictures appeared, the monkeys chose one of them by touching it, after which a computer delivered food on some percentage of the trials, depending on the picture picked. At first, the choice of picture C (the gray line in Fig. 8.1b) paid off about 60% of the time, on average; choosing picture A (the pale blue line) produced food a little more than 25% of the time; and choosing picture B (the pink line) never paid off. By trial and error, the monkeys learned to choose the picture with the highest pay-off percentage, picture C, more often than the other pictures. In an experimental twist, the pay-off percentages changed over the series of 300 trials. As we indicate with the pink line in Figure 8.1b, after 150 trials picture B became more valuable than picture C, and it became the best option a few trials later. By periodically exploring alternatives, the monkeys eventually noticed that picture B had begun to pay off more frequently, so they chose it more often.

A detailed analysis of the monkeys' behavior showed that they improved their chances of getting a pay-off by using the memory of a prior event, including a certain choice and a specific outcome. To understand this conclusion, imagine yourself in the monkey's shoes (despite the fact that monkeys don't wear shoes). From time to time over the first 150 trials, you choose picture B, but it never pays off, so you assign it a low value and don't choose it very often. All the same, you choose it on occasion just to make sure that you aren't missing out on something good, and on the 155th trial it pays off. Like the pay-by-paragraph scheme described earlier, one pay-off does not change the average value of picture B much, so you return to choosing the other pictures. A few trials later, though, the option you think is best isn't paying off at a rate you expect, so it's time to explore alternatives again. You remember that your choice of picture B had led to success in a specific instance in the past (a discrete goal-related event), so you choose picture B again despite its low average value. That's a good choice because its current value is much higher than its average value.

Monkeys need their ventral prefrontal cortex to improve their choices this way. A surgical removal of this brain area caused monkeys to take much longer than normal to switch their choice to picture B once it became the most valuable option. After picture B's value surpassed that of picture C (on the 150th trial), it took

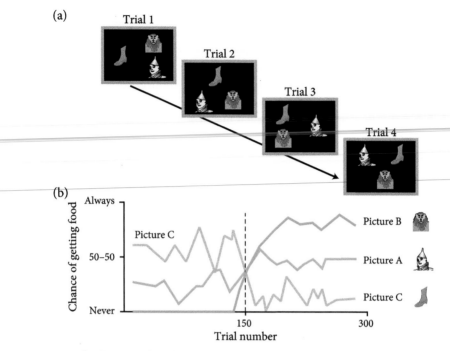

Fig. 8.1 The three-arm bandit task. (a) A series of four video screens viewed by monkeys. Three pictures appeared on each trial, and the monkey could only choose one. (b) The pink line shows the monkey's chance of getting food after the choice of picture B. Over the course of 300 trials, this chance changed: from no chance at all on the first 140 trials or so, to a very high chance during the final 100 trials. As the pay-off chance for picture B (pink line) increased, that for picture C (gray line) decreased, with the pay-off probability for picture A remaining somewhere in-between.

another 60–70 trials until these monkeys—the ones without their ventral prefrontal cortex—chose picture B (the best option) more often than they would by guessing (which is about 33% of the time). By the end of the 300-trial testing session, they chose the best option only about 50% of the time. Normal monkeys did much better. They took only about 20 trials to make the switch, and by the end of the session they chose the best option about 80% of the time.[19]

These findings reveal what the ventral prefrontal cortex does when it's intact and functioning. It uses the new memories that evolved in anthropoids—of discrete goal-related events—to speed learning and reduce errors. The use of these memories enables anthropoids to improve on the choices driven by their evolutionarily older, average-based memories. More specifically, the ventral prefrontal cortex provides a new way to update estimates of resource availability (as opposed to desirability) in guiding foraging choices.

It's tempting to assume that as new forms of memory evolved, they replaced older ones, but it's better to think in terms of augmentation. As anthropoids evolved, their new memories helped them adapt to volatile resources by supporting better (and ultimately safer) foraging choices. After scientists remove the ventral prefrontal cortex, monkeys revert to using older, average-based forms of memory, alone, so they learn more slowly and make more errors. In effect, they behave something like mammalian species that lack the new prefrontal areas that evolved in anthropoid primates.

Experts in brain research sometimes get uneasy with discussions about "new areas" in the brain; analyses of overall brain size or the number of neurons seem safer somehow. At heart, however, new cortical areas are just new places in the brain that combine information from previously existing areas in an innovative way and then send these representations to each other and to pre-existing areas. As new areas become integrated into a variety of large-scale cortical networks, their novel representations change the old areas in ways large and small, sometimes with profound consequences. In Chapter 10,* we'll explain how certain novel representations in the prefrontal cortex changed the hippocampus and, through that change, paved the road to human memory.

Remembering rules

The results of the "three-arm bandit" experiment are important, but the prefrontal cortex can help anthropoids do even better. In the next three sections, we'll consider how anthropoids avoid errors after fewer experiences—and sometimes only one.

Imagine playing a game that requires you to win 50 coins to get a smartphone. You begin by choosing between a blue box and a black box, with no hint about which one contains a coin (only one does). Then, after three choices, the boxes change. Instead of a blue box and a black box, you need to choose between a red box and a green box. After another three guesses, the options change again, this time to an orange box versus a yellow box—and so forth until the masters of the game run out of colors. (Crayola® makes 120 different colors of crayons, so this game could go on for a while.)

Now imagine that you don't pay any attention to color. Instead, you rely solely on averages: half of the boxes contain a coin and half don't. This strategy will yield 50 coins after about 100 choices, with some chance variation.

The next time you play the game, you notice that, for each pair of boxes, a box of one color always contains a coin and the alternative never does. It takes a few

---

* In the section entitled "Me, myself, and I."

pairs of boxes to discern this rule, but afterward the game becomes much easier. After the first guess for a given pair of boxes, you can get a coin for sure on the two subsequent choices. If the first guess wins a coin, then you simply choose the same color the next two times; if the first guess fails, then you shift to the other color and collect coins on the next two choices. Once implemented, this "color rule" yields an average of 2½ coins for every 3 choices.

Using an average-based strategy, you'd make about 50 errors on the way to collecting 50 coins. By employing the color rule, you'd only make 12–15 errors or so, depending on how quickly you discovered the rule. Because you'd get a smartphone in either case, this difference might not seem important. But now imagine that through time travel you're playing the same game in ancient Rome. You wouldn't get a phone, of course, but you'd want the coins. What if, after each error, you had a 1-in-1000 chance of being fed to the lions (or tigers or bears) in the Colosseum? In that case, the prospect of making 12–15 errors seems a lot safer than making 50 errors. Either way, you'd probably survive, but you'd stand a better chance if you used the color rule.

In an experiment like that game, monkeys faced a choice between two novel objects: one correct, the other incorrect. Instead of a "color rule" there was an "object rule." Every choice of a correct object produced a food reward; the alternative never did. If their first guess was successful—due to luck—the monkeys kept that object in memory and chose it again on the next trial. Otherwise, they switched to the alternative and kept that object in memory. Occasionally they forgot which object to choose, but even so the monkeys chose correctly about 90% of the time on the second trial for a given pair of objects, way more than the 50% they would get by using averages, alone.

To perform the task at this level of efficiency, the prefrontal cortex needs to receive visual information from the inferior temporal cortex. A particular surgical procedure* can block these interactions, however, and after it does, monkeys revert to the average-based approach. Accordingly, they make many more errors than normal monkeys do.[20] For each new pair of objects, they randomly chose between the objects on their second choice regardless of whether their first guess had paid off.

Although our anthropoid ancestors didn't bow down to a mighty emperor who might feed them to the lions, they risked predation every time they set out on an expedition for food. Their new prefrontal areas provided them with an important evolutionary advantage by reducing errors and thereby reducing risks. In the wild, every bad foraging choice led to a risky excursion for poor returns, so any reduction in cumulative errors improved their chances of survival.

---

* Scientists removed the prefrontal cortex from one hemisphere and the inferior temporal cortex from the other hemisphere. Along with cutting connections between the two hemispheres, this sugery eliminated direct interactions between the prefrontal and inferior temporal areas.

Two additional experiments revealed something about how the prefrontal cortex reduces errors. The first involved colored letters superimposed on a complex background scene; the second used colored letters that told a monkey what to do.

Remembering scenes

We mentioned earlier that context guides your choices. In that example, we pointed out that when you notice your phone vibrate, you'll behave differently at home than in a theater. The *letters-in-a-scene task* examines the influence of context on learning.

This task uses two colored letters in different locations on a complex background scene, presented on a video monitor. To a monkey, of course, these stimuli are just colored shapes. One of the letters is the correct choice, which the monkey indicates by touching it. For some reason, presumably because a background scene provides a visual context that helps the monkeys remember which letter is correct, monkeys learn the correct choice faster when the letters appear on a background scene than when they appear on a blank screen. This task might seem irrelevant to life outside the laboratory, but it draws on representations that enable anthropoids to choose foraging goals based on complex patterns of shapes and colors viewed at a distance.

In an experiment using this task, two groups of monkeys saw a series of 20 unique scenes, each with a different pair of colored letters superimposed on it. Scientists presented the same 20 scenes, in the same order, several times. A group of normal monkeys cut their error rate in half by the third presentation of a given scene and its two letters. This speedy learning was all the more impressive because 19 scenes (and choices) intervened before any particular scene (and its two letters) appeared again. After researchers removed the prefrontal cortex from a different group of monkeys, they made choices that were no better than a straight guess—a 50% error rate—by the time that they saw a particular scene and its letters for the third time.[20] By the eighth presentation, they had only reduced their error rate from 50% to 45%: not much of an improvement. We know that they remembered the task rule because they always chose one of the two letters, as opposed to touching somewhere else on the video monitor. Normal monkeys, in contrast, reduced their error rate from 50% to 5% by the eighth presentation of a scene.

Remembering instructions

The previous experiment involved learning to choose one colored letter instead of another. In the next experiment, monkeys learned to follow a symbolic instruction consisting of two, superimposed colored letters, which we'll call a cue.

Each cue appeared in the middle of a video screen, and the monkeys used trial and error to learn whether it instructed them to touch a square at the top, left, or right of the screen.[21] At first, the monkeys learned these relationships slowly, at about the same rate as rats facing the same problem. As they accumulated experience in performing this task, the monkeys learned faster and faster, eventually learning which location to choose the second or third time they saw a new cue. Rats never learn novel associations of this kind so quickly.

After a removal of their orbital and ventral prefrontal cortex, monkeys made a lot of errors, many more than normal monkeys do. Even after dozens of trials, they made the same percentage of errors as they would by guessing. They eventually learned which location to choose, but it took several days of practice and hundreds of trials to learn something that normal, experienced monkeys mastered in a few trials—and often only one.[21] In effect, the removal of their orbital and ventral prefrontal areas* produced an error rate in monkeys typical of that seen in rats.

Along with learning to associate a cue with a place, monkeys also learned to reduce their error rate by adopting a clever strategy. Sometimes, the cue from one trial repeated on the next one. When that happened, and if the previous choice had paid off, the monkeys simply repeated the choice they had just made. If, instead, the cue changed from the previous trial, they avoided their previous choice. In this way, monkeys could reduce their error rate even before they had a chance to learn what place each cue instructed them to touch, which was especially valuable before they reached their peak learning speed. Removal of the ventral and orbital areas completely abolished this strategy.[21]

The previous four experiments explored qualitative features of vision, such as color and shape. The next experiment examined a quantitative feature: order.

Remembering order

The *delayed-response task* reveals something important about the representation of order in the prefrontal cortex. This task is something like the game with colored boxes that we described earlier. Imagine that, in this game, you need to choose between two translucent white boxes. At the beginning of each trial, a light inside one of the boxes brightens it briefly. After a *delay period* of several seconds, both boxes brighten and remain bright. To get a coin, you then need to choose the box that had brightened briefly at the beginning of the trial.

The form of memory used to solve this task is often called *working memory*, and for decades the idea that the prefrontal cortex specializes in working memory

---

* In a different experiment, which used the same kind of task, the results indicated that the ventral prefrontal cortex makes the crucial contribution.

dominated research in this field. Scientists later learned that areas outside the prefrontal cortex also support the successful performance of working memory tasks.[22-24] Furthermore, the prefrontal cortex supports good performance on many tasks that don't rely on working memory at all.[25] In fact, most neurons in the prefrontal cortex represent a location that a monkey attends to in the present rather than a location remembered from the past.[26]

An often-neglected aspect of the delayed-response task involves the order of cues.[2] As monkeys perform this task, they face a long sequence of trials, and it's easy to confuse the cue on one trial with that on another. To understand why, imagine playing the lit-box game again. Instead of a single brief brightening, this time a series of brief brightenings occur for the two boxes. The order of flashes might be: left, right, right, left, right, left. Which cue designates the box with a coin? According to the task rule, only the last cue counts, but you don't know this rule at first. Like the colored-box game, you might just guess and get a coin half the time. But you'd eventually notice that the order of flashes matters and that the box cued last in a sequence of flashes always has the coin.

After gaining some experience with the delayed-response task, monkeys perform it well. After removal of their dorsolateral prefrontal cortex, however, they revert to guessing.[2] Either they can't remember a task rule based on order or they can't apply it. In either case, they can't use the rule to generate a plan of action and remember that plan until the time comes to make a choice. In support of this conclusion, we know that neurons in this area represent abstract sequences[27] and that nearby neurons represent the most recently cued location in a sequence.[28] Other nearby neurons represent the combination of a particular picture and its order in a sequence.[29]

Remembering this section

The prefrontal cortex helps anthropoids reduce errors in all of these tasks (and others). In some, it does so based on the memory of a goal-related event, which the anthropoid brain can establish after just one or a few experiences. In others, it reduces errors by applying clever strategies to novel problems.

Representations in action

Simply representing goal-related events does nothing to limit the number of risky excursions: not on its own. To gain this benefit, the anthropoid prefrontal cortex needs to influence actions, and to do that it has to send signals elsewhere in the brain. Earlier, we said that the prefrontal cortex specializes in representations that combine a sensory context with a goal, an outcome, and an action. Each component makes a distinctive contribution:

- The *sensory context* component applies a biasing signal to representations in "sensory areas" of cortex, which promotes the processing of some kinds of sensations at the expense of others. This function, called *top-down attention*, keeps the brain from becoming overloaded, as we discussed in Chapter 3,* and it does so in a highly selective way. If a particular sensation is associated with a good foraging choice, it makes sense to be on the lookout for it at the expense of other sensations.
- The *goal* component enables the representation of plans and aspirations on a variety of time horizons. A concrete, immediate goal might involve the acquisition of a particularly sweet-looking berry while at a pick-your-own farm; an abstract, long-term goal might be to photograph a pride of lions in the Serengeti, perhaps decades in the future.
- The *outcome* component stores the consequences of a choice. It also enables anthropoids to take into account their current biological needs, such as states of selective satiation, as they decide on a foraging journey.
- The *action* component applies a biasing signal to competing movement representations to achieve the current goal.

Anthropoid augmentations reconsidered

As new prefrontal areas evolved in anthropoids, their representations established novel forms of memory. Older forms of memory—such as Pavlovian and instrumental associations—served animals well through most of evolutionary history, but when evolving anthropoids faced a heightened degree of volatility in their preferred foods, they had to risk predation (and waste foraging effort) for lesser returns. Because they lived for a long time, depended on fruit for nourishment, journeyed long distances in daylight, and faced the threat of predation every time they did, our anthropoid ancestors benefited from any reduction in the number of food-gathering expeditions that they needed to make. As the rules of the (foraging) game changed, they needed to learn quickly. They could do so, and thereby gain an evolutionary advantage, because of new prefrontal areas that brought together information about sensory contexts, goals, actions, and outcomes in a unique way. As a result, anthropoids survived and thrived in a rapidly changing world teeming with predators.

In Figure 8.2, we place labels on an anthropoid brain to indicate some of the new representations that emerged during anthropoid evolution. In Chapter 7, we

---

* In the section entitled "Advantages of instrumental memory."

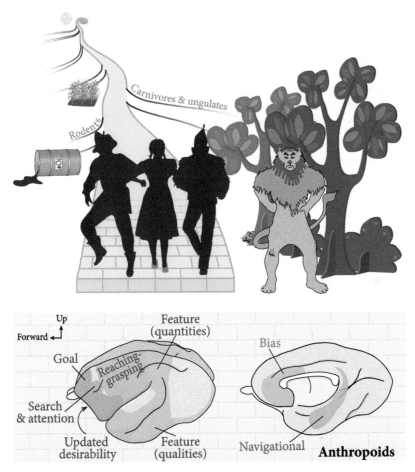

Fig. 8.2 Some anthropoid innovations. The representations that emerged or became elaborated during anthropoid evolution are labeled on an anthropoid brain: in this case, the brain of a titi monkey. Updated valuation representations reside in the orbital cortex, which is hidden from view (as indicated by the curved arrow toward the lower left). The Cowardly Lion's addition to Dorothy's band of brothers is analogous to the new anthropoid representations that joined those inherited from more-distant ancestors. In *The Wizard of Oz*, the King of the Forest proves his prowess as "a lion not a mou-ess." (Pay no attention to the mammal behind the tree.) In the background, we depict icons from the past: a compass device marks the beginning of the evolutionary road to human memory; the cornfield commemorates Dorothy's encounter with the Scarecrow; and an oil can represents the addition of the Tin Man to her entourage.

considered the novel *feature representations* of anthropoids, located mainly in the parietal and temporal lobes; in this chapter we've highlighted the *goal representations* of the prefrontal cortex.

This chapter began with the Scarecrow's fear of wild animals: lions, and tigers, and bears. Some animals are aggressive and dangerous; others peaceful and friendly. In the next chapter, we'll explore how evolution enabled you to categorize animals as dangerous or friendly and how this capacity, in turn, established the foundation for human culture.

References

1. Cheney, D. L. & Seyfarth, R. M. *How Monkeys See the World: Inside the Mind of Another Species* (Chicago, IL: University of Chicago Press, 1990).
2. Passingham, R. E. & Wise, S. P. *The Neurobiology of the Prefrontal Cortex: Anatomy, Evolution, and the Origin of Insight* (Oxford, UK: Oxford University Press, 2012).
3. Williams, B. A., Kay, R. F., & Kirk, E. C. New perspectives on anthropoid origins. *Proceedings of the National Academy of Sciences U.S.A.* 107, 4797–4804 (2010).
4. Zuberbühler, K. & Janmaat, K. in *Primate Neuroethology*, pp. 64–83 (eds M. L. Platt & A. A. Ghazanfar) (Oxford, UK: Oxford University Press, 2010).
5. Teuber, H. L. Neurophysiology, effects of focal brain lesions. *Neuroscience Research Progress Bulletin* 10, 381–384 (1972).
6. Teuber, H. L. Unity and diversity of frontal lobe function. *Acta Neurobiologiae Experimentalis* 32, 615–656 (1972).
7. Miller, E. K. & Cohen, J. D. An integrative theory of prefrontal cortex function. *Annual Reviews of Neuroscience* 24, 167–202 (2001).
8. Kim, J.-N. & Shadlen, M. N. Neural correlates of a decision in the dorsolateral prefrontal cortex of the macaque. *Nature Neuroscience* 2, 176–185 (1999).
9. Genovesio, A., Brasted, P. J., Mitz, A. R., & Wise, S. P. Prefrontal cortex activity related to abstract response strategies. *Neuron* 47, 307–320 (2005).
10. Barraclough, D. J., Conroy, M. L., & Lee, D. Prefrontal cortex and decision making in a mixed-strategy game. *Nature Neuroscience* 7, 404–410 (2004).
11. Tsujimoto, S. & Sawaguchi, T. Neuronal representation of response-outcome in the primate prefrontal cortex. *Cerebral Cortex* 14, 47–55 (2004).
12. Hayden, B. Y. & Platt, M. L. Neurons in anterior cingulate cortex multiplex information about reward and action. *Journal of Neuroscience* 30, 3339–3346 (2010).
13. Wallis, J. D. & Miller, E. K. Neuronal activity in primate dorsolateral and orbital prefrontal cortex during performance of a reward preference task. *European Journal of Neuroscience* 18, 2069–2081 (2003).
14. Kennerley, S. W. & Wallis, J. D. Evaluating choices by single neurons in the frontal lobe: outcome value encoded across multiple decision variables. *European Journal of Neuroscience i* 29, 2061–2073 (2009).
15. Amiez, C. & Petrides, M. Selective involvement of the mid-dorsolateral prefrontal cortex in the coding of the serial order of visual stimuli in working memory. *Proceedings of the National Academy of Sciences U.S.A.* 104, 13786–13791 (2007).

16. *USA Today* article: Why India is going bananas over birth control for monkeys. Available at: https://www.usatoday.com/story/news/world/2017/05/11/monkeys-india-pests-contraceptives/101466580 (Gannett, 2017).
17. *BBCNews* article: Monkeys attack Delhi politician. Available at: http://news.bbc.co.uk/2/hi/south_asia/7055625.stm (London, UK: British Broadcasting Corporation, 2007).
18. Walton, M. E., Behrens, T. E., Buckley, M. J., Rudebeck, P. H., & Rushworth, M. F. Separable learning systems in the macaque brain and the role of orbitofrontal cortex in contingent learning. *Neuron* **65**, 927–939 (2010).
19. Rudebeck, P. H., Saunders, R. C., Lundgren, D. A., & Murray, E. A. Specialized representations of value in the orbital and ventrolateral prefrontal cortex: desirability versus availability of outcomes. *Neuron* **95**, 1208–1220 (2017).
20. Wilson, C. R. E., Gaffan, D., Browning, P. G. F., & Baxter, M. G. Functional localization within the prefrontal cortex: missing the forest for the trees? *Trends in Neurosciences* **33**, 533–540 (2010).
21. Bussey, T. J., Wise, S. P., & Murray, E. A. The role of ventral and orbital prefrontal cortex in conditional visuomotor learning and strategy use in rhesus monkeys (*Macaca mulatta*). *Behavioral Neuroscience* **115**, 971–982 (2001).
22. Meyer, T. & Rust, N. Single-exposure visual memory judgments are reflected in inferotemporal cortex. *Elife* **7**, 32259 (2018).
23. Dotson, N. M., Hoffman, S. J., Goodell, B., & Gray, C. M. Feature-based visual short-term memory is widely distributed and hierarchically organized. *Neuron* **99**, 215–226 e214 (2018).
24. Rademaker, R. L., Chunharas, C., & Serences, J. T. Simultaneous representation of sensory and mnemonic information in human visual cortex. *BioRxiv*, doi: 10.1101/339200 (2018).
25. Wise, S. P. Forward frontal fields: phylogeny and fundamental function. *Trends in Neurosciences* **31**, 599–608 (2008).
26. Lebedev, M. A., Messinger, A., Kralik, J. D., & Wise, S. P. Representation of attended versus remembered locations in prefrontal cortex. *Public Library of Science (PLoS) Biology* **2**, 1919–1935 (2004).
27. Shima, K., Isoda, M., Mushiake, H., & Tanji, J. Categorization of behavioural sequences in the prefrontal cortex. *Nature* **445**, 315–318 (2007).
28. Konecky, R. O., Smith, M. A., & Olson, C. R. Monkey prefrontal neurons during Sternberg task performance: full contents of working memory or most recent item? *Journal of Neurophysiology* **117**, 2269–2281 (2017).
29. Naya, Y., Chen, H., Yang, C., & Suzuki, W. A. Contributions of primate prefrontal cortex and medial temporal lobe to temporal-order memory. *Proceedings of the National Academy of Sciences U.S.A.* **114**, 13555–13560 (2017).

9

Human heritage

Specialization for generalization

"If I asked you where the hell we were," said Arthur weakly, "would I regret it?"

Ford stood up. "We're safe," he said.

"Oh good," said Arthur.

"We're in a small galley cabin," said Ford, "in one of the spaceships of the Vogon Constructor Fleet."

"Ah," said Arthur, "this is obviously some strange use of the word *safe* that I wasn't previously aware of."

—*The Hitchhiker's Guide to the Galaxy*, Douglas Adams

The word "safe," like other words, means whatever two or more people agree it means. An alternative that some people seem to believe, which is that dictionaries dictate the meaning of words, can't be right because people prattled along for millennia before the first dictionary. All the same, words work best when they have widely agreed meanings. To some, the words "Ford Prefect" bring to mind a British automobile from the 1950s. So, it makes perfect sense to use "Ford Prefect" as a humorous name, as Douglas Adams does in *The Hitchhiker's Guide to the Galaxy*. Few Americans get the joke, but so what? All readers eventually learn that Ford Prefect is a visitor to Earth from the planet Betelgeuse, who saves the earthling Arthur Dent moments before the Vogons obliterate our planet.

We'll return to *The Hitchhiker's Guide* at the end of this chapter, but first, after a few points about human evolution, we'll address two evolutionary changes that shaped the earthling brain. One, which we'll discuss at length, involves what experts call *semantic memory*. Unfortunately, this label doesn't capture the underlying concept very well. The word "semantic" usually refers to the meanings of words or other symbols. But semantic memories also encompass concepts, knowledge of categories, and other generalizations about the world. Our newly minted term, *cultural memory*, covers all of this and more—at every level of complexity. The second evolutionary change, which we'll touch on only briefly, involves another kind of generalization: the capacity to discern meaningful relations in the world.

The Evolutionary Road to Human Memory. Elisabeth A. Murray, Steven P. Wise, Mary K. L. Baldwin and Kim S. Graham, Oxford University Press (2020). © Oxford University Press.
DOI: 10.1093/oso/9780198828051.001.0001

(a)   (b)

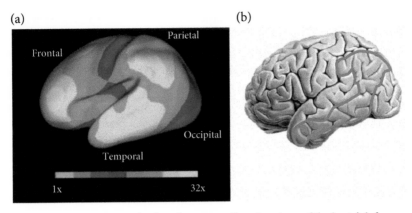

Fig. 9.1 Brain expansion and cultural memory. Two drawings of the brain's left hemisphere, with the front of the brain to the left. (a) The color code shows the degree of regional cortical expansion. (b) The pink areas play a role in cultural (semantic) memory. The brain on the left is rendered as if the folds of the brain are flattened out, while the brain on the right depicts the folds as they are.

Part a: Adapted from Jason Hill, Terrie Inder, Jeffrey Neil, Donna Dierker, John Harwell, and David Van Essen, Similar Patterns of Cortical Expansion during Human Development and Evolution, Proceedings of the National Academy of Sciences of the United States of America, 107 (29), pp. 13135–13140, Figure 7a, doi.org/10.1073/pnas.1001229107 © National Academy of Sciences, 2010.

Part b: Adapted from Jeffrey R. Binder, Rutvik H. Desai, William W. Graves, and Lisa L. Conant, Where Is the Semantic System? A Critical Review and Meta-Analysis of 120 Functional Neuroimaging Studies, *Cerebral Cortex*, 19 (12), pp. 2767–2796, Figure 7a, doi: 10.1093/cercor/bhp055, Copyright © 2009, Oxford University Press.

CHAPs, not chimps

About 5–7 million years ago, an ape species split into two branches. We call this species the *CHAP*, an acronym for *chimpanzee–human ancestral population*,[*] and in Figure 2.4 we illustrate its location on our family tree. The CHAPs were not chimps, and they certainly were not humans. In fact, no one knows how closely they resembled any modern species.[1] From this common origin, one group of CHAPs eventually evolved into modern chimpanzees and bonobos; the other into hominins, including modern humans. In Chapter 10,[†] we'll discuss some of the cultural and cognitive developments that occurred during hominin evolution; in this chapter, we'll focus on brain anatomy.

In Figure 9.1a, a color code identifies the parts of neocortex that expanded the most during ape and human evolution[2]: some changes occurred before the time of the CHAPs and some afterward. Yellow regions expanded up to more than 30-fold relative to the last common ancestor of humans and Old World monkeys; blue areas expanded much less, if at all. Note that the yellow regions encompass

[*] A term coined by Stephanie Keep of the National Center for Science Education.
[†] In the section entitled "Explosions."

much of the frontal, temporal, and parietal lobes, but only parts of each one. In Chapters 7 and 8, we said that frontal, temporal, and parietal lobes each developed new specializations during the course of anthropoid evolution. Parietal areas represent quantitative aspects of the visual world, such as numbers, distances, durations, and order; temporal areas represent qualitative aspects of the world, such as various combinations of shapes, colors, and surface appearance, along with some analogous properties for sounds. The prefrontal cortex receives this information and uses it to generate goals—based on the memory of discrete goal-related events, problem-solving strategies, and abstract behavioral rules. These memories speed learning and reduce errors, thus increasing the likelihood of surviving and producing progeny. Two points are especially important for this chapter:

- For the parietal cortex, *relative* or *relational* quantities come to the fore: the number of berries will be greater in one place than another, and one group of berries will be farther than another, which will take longer to reach.
- For the temporal cortex and its qualitative representations, categorization and generalization are especially important. For ancient anthropoids with a keen eye for dangerous hyenas, the appearance of a never-before-seen animal with similar-looking features would merit the same degree of caution.

In this chapter, we'll advance the idea that new forms of generalization emerged as hominin brains enlarged: one derived from quantitative features represented in the parietal lobe; the other from qualitative features represented in the temporal lobe. Bear in mind, however, that all animals generalize to an extent, and some theorists think that this is the fundamental function of the neocortex in all mammals.[3]

From foraging to culture

A horse is a horse

As Douglas Adams might have put it in *The Hitchhiker's Guide*: cultural memories underlie your knowledge about life, the universe, and everything. They also empower you to establish broad and abstract generalizations, to generate predictive inferences, and to acquire and update knowledge about your place in the world and society. Unlike *participatory* (episodic) memories, which record what happened at a particular time and place, *cultural* memories rarely depend on when or where you encountered a particular piece of information.

A simple example illustrates this point. In Chapter 2, we mentioned that people often regard pigs as greedy animals and apply that generalization to anyone who takes more than their fair share of something. Everyone understands the image of politicians "feeding at the trough of their corporate benefactors." This saying does not imply that politicians walk on four legs or have curly tails, snouts, and hooves. To humans, the concept "pig"* goes beyond the features that belong to genuine pigs. Once established, people can use a cultural construct like "pig" very flexibly, well beyond the limits of the observable world:

> Question: What is the difference between a politician and a flying pig?
> Answer: The latter includes the letter "f."

For animals that have been so beneficial to so many people for so long,[4] it's regrettable that the word "pig" serves as an insult. Like most cultural knowledge, the concept of a "lying pig" does not depend on any particular event; no pig has ever lied to anyone. It's a context-free generalization, one that pertains to an entire category of individuals—in this case, politicians. Like many generalizations, it might be unfair: certainly to pigs and possibly to politicians, as well.

An important characteristic of cultural memory is that different kinds of sensations can elicit the same memory. People think of a horse whether they see a picture of one, listen to someone say the word horse, hear neighing or nickering sounds, or—for Americans of a certain age—when they encounter the name *Mister Ed*. All of these sensations elicit the same representation: "horse." In addition to storing knowledge, such as the fact that Mister Ed was a talking horse in a 1960s television series, cultural memory enables predictive inferences. Without any firsthand knowledge, we can infer that a politician who depends on campaign contributions from the oil and gas industry might not be entirely objective when evaluating the scientific evidence for global warming. Likewise, in Figure 7.1 we illustrate something we know will never happen: a pig winning the Kentucky Derby. Without going through the trouble of staging such a race, everyone recognizes it as a fantasy because of the features that compose the concept "pig" as opposed to the concept "horse." The latter comes along with a certain feature—sustained, speedy galloping on long, graceful legs—while "pig" brings to mind a slower prance-like form of locomotion propelled by stubby limbs.

Brain evolution provides some clues about how cultural knowledge arose. In Figure 9.1, we highlight the similarity between cortical areas thought to underlie cultural memories (Figure 9.1b, in pink)[5, 6] and those that expanded dramatically during ape and hominin evolution (Figure 9.1a, in yellow).[7,8] Brain-imaging studies[9] have revealed enhanced cortical activation during tasks that involve cultural memories, typically in the pink areas of Figure 9.1b. The similarity between these two brain maps might be a coincidence, but it seems more likely that at least

---

* In this chapter, we use quotation marks to mark a concept or generalization such as "pig" or "horse."

some of the yellow cortical areas in Figure 9.1a expanded in the service of cultural memory.

Four-legged ducks

Like most people, you probably take your cultural memories for granted. When you see a pig, you know what it is, instantly and effortlessly. Imagine, though, waking up one day without that knowledge. When you next saw a pig, you'd have no idea what it was. That, alone, might be a small loss, but what if many of your other memories vanished, too? This is what happens in a brain disease called *semantic dementia*, which results from the progressive withering of certain parts of the brain. The study of patients with this disorder has produced much of our knowledge about cultural (semantic) memory.

A patient called J. L., a 60-year-old businessman, typifies this disorder.[10] He had:

a two-year history of difficulty in remembering the names of people and places. More recently his vocabulary had diminished causing word finding difficulty in conversation. The family had noted [his] problems understanding the meaning of words that had previously been well within his vocabulary, for instance the names of foods. He was no longer familiar with the names of the local villages in the region where he had lived all his life . . . [The disease] also affected his ability to recognise people by face, name, or description. He had also exhibited problems identifying real objects and animals, for instance he had been frightened by finding a snail in the back yard and thought that a goat was a strange creature. Despite these difficulties, his day-to-day and autobiographical memory remained . . . very good.

As patient J. L. illustrates, semantic dementia causes a highly selective loss of cultural memory. Because he failed to recognize a snail, he found it frightening to come across this "new" animal in his backyard. But he didn't forget everything, at least not initially; he still knew about animals in general, and he knew that they differed from inanimate objects. The selectivity of J. L.'s memory loss is typical of such patients. As the brain damage progresses in this disease, patients lose their cultural memories bit-by-bit. Initially, they might be unable to select a domestic pig from pictures of other pig-like animals, such as peccaries, babirusas, and warthogs. But they could still select a pig from a lineup that included elephants, echidnas, and eagles. Later, they might lose their ability to do that, but still be able to pick a pig from a lineup that included saws, screwdrivers, and shovels.

The selectivity of memory loss in these patients has another aspect, too. Some kinds of memory seem immune to the disease, for the most part. Patient J. L., for example, could remember events that he had experienced, especially more recent ones. This is what the clinical investigators meant when they wrote that "his day-to-day and autobiographical memory remained . . . very good." In general, patients with semantic dementia can also understand numbers and other quantitative information,[11] and they retain most of their physical skills, including their ability to play the piano, drive a car, and use tools, although they might not be able to name them.[12] (In Chapter 6,* we discussed patients with the opposite problem; they could remember the names of tools but not how to use them.)

The memory loss in semantic dementia is also selective in an additional way. Rarely encountered and unusual items deteriorate the most and earliest in the course of the disease. These vulnerabilities identify some important characteristics of cultural memory:

- Frequently encountered items form strong representations, both through repeated experience with them (repetition) and neural processes called consolidation and reconsolidation, which strengthen memories (after a period of vulnerability). People encounter dogs quite often, so the memory "dog" becomes strongly established. Elephants come to mind less frequently, and aardvarks come up only rarely. So, these representations never get as strong or stable as the memory "dog." As a result, patients with semantic dementia will forget "aardvark" very early in the disease, and they will forget "elephant" long before they forget "dog."
- Unusual items—those included in a category but atypical of it—are vulnerable for a different reason. By definition, they have a low number of features in common with other items in their category. Memories that share more features tend to reinforce each other, so the memories of typical items are stronger than those of atypical ones. For example, a dog is typical of the category "mammal," but a duck-billed platypus is so atypical that when the first specimens arrived in England from Australia, the scientists examining them suspected a hoax. Patients with semantic dementia will forget "platypus" before "dog" because dogs have features typical of mammals: four straight legs; a soft, wet nose; and live birth, for example. No platypus has any of these traits. (People also encounter the concept of a "platypus" infrequently, so it has vulnerabilities based on both frequency and typicality.)

As they lose cultural memories in these selective ways, even the strongest (most frequent and typical) representations become vulnerable. When seeing a horse, for example, patients with semantic dementia might resort to calling it an animal rather than a horse, or they might call it a dog. If asked, they know that "horse" is a word, but it no longer activates the representation "horse" in their brain. So, they

---

* In the section entitled "Affordances."

no longer remember the features included in the generalization "horse," such as the ability to run at high speeds with stamina. Although these patients once had the representation "horse" in their brains, the disease caused it to deteriorate. Unfortunately, the disease is progressive; once lost, these memories rarely return.

The loss of cultural memories has wide-ranging consequences. In addition to being frightened by strange "new" animals they previously recognized as run-of-the-mill denizens of their backyard gardens, these patients might fail to remember the characteristic features of items.[13] For example, they might forget that carrots are usually orange or that elephants have big ears and long trunks. In an experiment that demonstrated this problem, scientists showed these patients pictures of both a typical carrot and a fictional green carrot. The researchers then asked them to pick out the "real" item. The patients usually chose the green carrot, even though they had normal color vision.[14] In a related experiment, patients saw two drawings of an elephant. One depicted a typical elephant with its mammoth ears and long trunk; the other replaced those features with small ears and stubby noses, which are more typical of mammals. The patients usually chose the small-eared, stubby-nosed elephant as representative of the real animal.[15] When asked to draw animals from memory, the patients made the same kinds of mistakes, leaving out the specific features that distinguish various animals from each other, such as big ears (for elephants), humps (for camels), and bipedal locomotion* (for ducks). They drew four-legged ducks, flat-backed camels, and elephants with small ears and no trunk.[16] In Figure 9.2, we present two examples in the bottom row.

These observations show that as their cultural memories deteriorated, patients with semantic dementia lost their memory of an item's characteristic features, starting with those infrequently encountered in a given category. They also forgot atypical members of a particular category. So, when performing the tasks described earlier, these patients relied on their general knowledge about plants, animals, and other items. Because the green of chlorophyll dominates the color of most plants, these patients most often chose green carrots instead of orange ones. Having lost their memory of carrots, the predominance of green in plants made the choice of a green carrot seem sensible (albeit inaccurate). Likewise, because these patients had lost their concepts of camels and ducks, and because the most commonly encountered animals have flat backs and four legs, they considered hump-free camels and four-legged ducks to be the most accurate versions of these animals. In general, the patients retained their memories of features shared by most items in a category but not uncommon or atypical features. Accordingly, they expected all plants to be green and all animals to have four legs, flat backs, and small ears.

---

* As Snowball, the pig, says in *Animal Farm*: "four legs good, two legs bad." What he means is that, in his porcine opinion, animals are better than humans. In reality, plenty of animals use bipedal locomotion. But quadrupedal locomotion is more typical, mostly because it's the ancestral form of locomotion in land vertebrates.

Four-legged duck   Flat-backed camel

Fig. 9.2 Four-legged ducks and flat-backed camels. A patient with semantic dementia viewed the drawing in the top row and tried to reproduce it later, from memory. The bottom row shows the result, a mixture of accurate memories (such as duck tails, camel tails, and duck bills) and inaccurate ones (such as four-legged ducks and hump-free camels).

Reprinted by permission from Springer Nature: *Nature Reviews Neuroscience*, 8 (12), Where Do You Know What You Know? The Representation of Semantic Knowledge in The Human Brain, Karalyn Patterson, Peter J. Nestor, and Timothy T. Rogers, pp. 976–987, Figure 2d, Copyright © 2007, Springer Nature.

The loss of cultural memories can be linked to the brain areas damaged by the disease. For reasons that remain poorly understood, a part of the temporal lobe toward the front of the brain degenerates earlier and faster than other areas. As the disease progresses, more and more of the temporal lobe withers away. Connections that lie beneath this part of the cortex also deteriorate, especially those that connect the frontal and temporal lobes.[17] This pattern of degeneration points to the front part of the temporal lobe as the region that pulls together the representations of cultural memory. Scientists call such areas *hubs*, which implies that they draw on information from several other parts of the brain.

The hubbub about hubs

According to the hub-and-spoke theory of cultural memory, most cortical areas represent specialized kinds of information and connect with a hub that extracts generalizations. In the mind's eye, this arrangement resembles a bicycle wheel, in which spokes connect a hub to a rim. In Figure 9.3, we depict the anatomical arrangement of selected spoke areas in blue, with the hub in pink.

The hub differs from its spoke areas in that its representations don't depend on whether items in a given category share many—or even any—physical features. The concept "mother" serves as an example. Human mothers throughout the

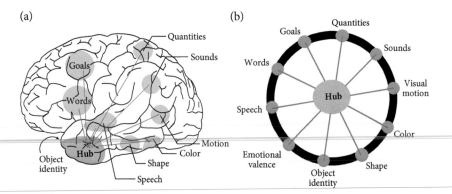

Fig. 9.3 The hub-and-spoke model. (a) The location of the hub for cultural memory (pink) and some selected spoke areas (blue) in the brain. (b) A figurative depiction of the hub-and-spoke model.

Left part reprinted from *Neuropsychologia*, 48 (5), Gorana Pobric, Elizabeth Jefferies, and Matthew A. Lambon Ralph, Amodal Semantic Representations Depend on Both Anterior Temporal Lobes: Evidence from Repetitive Transcranial Magnetic Stimulation, pp. 1336–42, doi.org/10.1016/ j.neuropsychologia.2009.12.036 Copyright © 2010 Elsevier Ltd., with permission from Elsevier.

world share many sensory features (including, quite often, an exasperated expression), but the cultural concept "mother" includes anything that produces smaller versions of something resembling itself. Obviously, a "mother ship," which aliens use to launch smaller attack vessels, doesn't share any sensory features with a human mother. Neither does a mother lode, a motherboard, or Mother Nature. Nevertheless, the concept "mother" applies to them all.

Results of brain-imaging experiments have provided support for the hub-and-spoke model. Cortical areas in the front part of the temporal lobe showed the same, high level of activation for particular cultural memories, regardless of whether words, pictures, or sounds elicited them.[18,19] This similarity is as expected for a hub area that binds separate pieces of information into generalizations, concepts, and categories regardless of the specific sensation that brings them to mind. Spoke areas toward the back of the temporal lobe showed different degrees of activation when a word elicited a memory as opposed to when a picture did so, as expected for specialized areas.

Because of the hub-and-spoke arrangement, damage to the hub eliminates cultural memories regardless of the sensations that elicit them. For example, when the memory "horse" is lost, neither neighing, nickering, pictures of a galloping horse, nor the word "horse"—written or spoken—can bring the appropriate memory to mind. When cortical damage spares the hub but impairs the functioning of one or more spoke areas, people lose specialized memories rather than generalizations. For example, localized damage to one of the spoke areas causes *word deafness*.

These patients aren't really deaf, but they can't recognize spoken words despite the fact that they can understand written words and pictures.[20] As we explained in Chapter 7,* localized damage to a different spoke area causes *prosopagnosia*, in which patients cannot recognize people by their facial features but know them by name or by voice. These examples illustrate a key point: in semantic dementia, damage to the hub area causes an impairment that is more general than either word deafness or prosopagnosia. Patients with this disease can't recognize people by either their faces, names, or voices, and the hub-and-spoke model explains why.

The hub-and-spoke model also explains something we mentioned earlier. Patients with semantic dementia tend to remember frequently encountered and typical items better than rarely encountered and atypical ones. A patient might forget "platypus," a rare and atypical mammal, while holding onto memories like "dog," "mammal," and "animal." To account for this selectivity, scientists distinguish among three levels of concepts: *general-level* concepts (such as "animal"); *basic-level* concepts (such as "dog"); and *specific-level* concepts (such as Toto and Cairn terrier).[21]

A series of experiments have demonstrated the importance of these distinctions. In one study, people saw a picture of a dog and had to answer a basic-level question: "Is this a dog?" They answered this basic-level question faster and more accurately than the general-level question: "Is this an animal?".[22] In general, healthy people recall basic-level concepts better and faster than either specific-level or general-level concepts. Unlike healthy people, patients with semantic dementia do best with general-level concepts. In one experiment, when these patients saw a picture of a kingfisher, they could identify it as an animal, a general-level concept, faster and more accurately than they could identify it as a bird, a basic-level concept.[22] When people sustain damage to their hub area, basic-level representations suffer the most. The evolution of a new cortical hub seems to have enabled this new level of generalization.

Dimming and distortion

Brain disorders other than semantic dementia can also cause impairments in cultural memory. Patients who contract a viral infection called *Herpes simplex* encephalitis have inflammation in their brain that especially affects their temporal lobe. Their impairment differs from the one seen in semantic dementia, and some experts have described the difference in terms of *dimming* versus *distorting* memories.[23] A "dimmed" memory, as observed in semantic dementia, is analogous to a

---

* In the section entitled "Distinctions from conjunctions."

picture that has faded away after years of exposure to direct sunlight, and it results from a long-term loss of neurons and connections in the cortical hub. In contrast, some forms of *Herpes simplex* encephalitis lead to "distorted" memories, which correspond to pictures that have been deformed. Image distortion makes people more likely to mistake one picture for another because both have become difficult to recognize, and the same goes for the distortion of cultural memories. Patients with encephalitis often retrieve the wrong cultural memory: "cow" when they see a horse, for example. Errors like this sometimes occur in semantic dementia, but not as often as in encephalitis.

Tempo and the temporal lobe

Neurosurgeons sometimes treat epilepsy by removing the cortical area that initiates seizures, as in Henry Molaison's case.* This procedure usually reduces the frequency of seizures, but it can have unwanted side effects. After removal of the front part of the temporal lobe, patients often have impairments in cultural memory.

In one test, patients with a partial removal of their temporal lobe took an unusually long time to decide whether a picture contained a bird.[24] Healthy people made decisions much faster than the patients and also made slightly fewer errors. Because the surgeons removed the hub from only one hemisphere, these patients had mild impairments compared to patients with semantic dementia,[25] but their problems involved the same kind of memory.

Scientists have confirmed this conclusion by studying healthy people.[26, 27] When they used electromagnetic stimulation to disrupt neural processing in the front part of the temporal lobe, people took longer to decipher the meaning of words or to remember the names of items in pictures. This observation shows that temporary interference with temporal lobe function results in a pattern of impairments that resembles patients with surgical removal of the same brain area.

In another experiment, which we illustrate in Figure 9.4a, people saw an array of three items: one at the top and two at the bottom. They had to choose which of the two items at the bottom goes with the item at the top: for either words or pictures. For example, a stool goes with a piano, but an armchair doesn't. During disruptive stimulation of the front part of the temporal lobe, people usually made the correct choice, but they did so more slowly than usual, as we illustrate in Figure 9.4b. Although this slowing amounted to only one or two tenths of a second, these results show that the front part of the temporal lobe needs to operate normally for the efficient use of cultural memories. Disruption of cortical areas

---

* We summarized Henry's surgery and its side effects in Chapter 1, in the section entitled "The first patient."

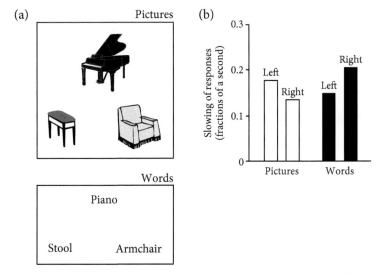

Fig. 9.4 A semantic association task. (a) Picture-based (top) and word-based (bottom) versions of a cultural memory test. For each version, people chose one of the two items at the bottom of the array based on the item at the top. (b) Disruptive stimulation slowed responses to arrays like the ones shown in (a). The taller the bar, the greater the slowing compared to an experimental condition without cortical stimulation. Scientists disrupted the temporal lobe hub separately in the left and right hemispheres, as labeled.

Reprinted from *Neuropsychologia*, 48 (5), Gorana Pobric, Elizabeth Jefferies, and Matthew A. Lambon Ralph, Amodal Semantic Representations Depend on Both Anterior Temporal Lobes: Evidence from Repetitive Transcranial Magnetic Stimulation, pp. 1336–42, doi.org/10.1016/j.neuropsychologia.2009.12.036 Copyright © 2010 Elsevier Ltd., with permission from Elsevier.

near the back of the brain didn't slow these choices at all, which confirms the importance of the front parts of the temporal lobe, the hub, for cultural memories. Disrupting the hub's activity didn't interfere with judgments about numbers or other quantitative information, presumably because the parietal lobe continued to function normally.

Cortex for culture

The cortical hub for cultural memory derives from parts of the temporal lobe that emerged or became elaborated during anthropoid evolution, as we discussed in Chapter 7. When they first emerged, these then-new cortical areas helped your distant anthropoid ancestors find food at a distance, and they also contributed to the complex social systems that developed long before the advent of apes or humans. During hominin evolution, the temporal lobe expanded, and as it did it developed

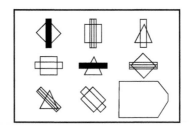

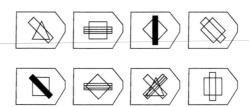

Fig. 9.5 A relational reasoning puzzle. People attempted to complete the nine-item array at the top by choosing one of the eight items in the bottom two rows. We'll reveal the correct choice somewhere in this chapter.

Reprinted from Carpenter P, Just M, Shell P, What One Intelligence Test Measures: A Theoretical Account of Processing in the Raven Progressive Matrices Test. *Psychological Review*, 97, 404–431. © 1990 American Psychological Association.

new and more generalized representations. In Chapter 10,* we'll suggest that the representations underlying cultural memory might have been the most recent in a series of evolutionary developments along the road to human memory.

From relations to reasoning

So far in this chapter, we've highlighted the role of the temporal lobe in representing generalizations. Parietal and frontal areas contribute to generalizations in a way that differs from the temporal lobe. Many of the relevant experiments involve puzzles that tax analogical or relational reasoning. The term *relational reasoning* refers to the ability to discern meaningful patterns among sensations, concepts, words, and quantities; *analogical reasoning* involves one such pattern: similarity.

In one set of experiments, people solved puzzles of the sort that we illustrate in Figure 9.5. They had to work out, by understanding the relations among the items in rows and columns, which of the eight options in the bottom two rows should be placed in the lower, right corner of the puzzle at the top. (We'll reveal the correct

---

* In the section entitled "The cross-domain cathedral."

answer at the end of this section.) As people solved this puzzle, the level of neural activation increased significantly in particular parts of the prefrontal and parietal cortex.[28] A related study examined the effects of practicing for a law school admission test that relied heavily on analogies. Over weeks of study, the same parietal and prefrontal areas developed stronger connections with each other, as revealed by coordinated levels of activation.[29]

In an influential line of research, John Duncan[30] pulled together several ideas about general problem-solving under the banner: *multiple-demand system*.* He used this term to refer to the ability to cope with a wide range of difficult problems, including novel ones, along with other kinds of high-level cognition. In several brain-imaging experiments, a set of parietal and prefrontal areas exhibited especially prominent increases in activation for many different kinds of cognitive challenges,[30] and patients with damage to these areas perform poorly on a diverse array of cognitively demanding tasks.[31]

Analogical reasoning enables established knowledge to provide insight into new problems, and a story about the Wright brothers exemplifies its power. For millennia, people could see that birds use wings to get airborne. So, when engineers first attempted to design heavier-than-air flying machines, they used their knowledge about bird wings to get started. One such pioneer, Otto Lilienthal, designed a bird-like wing, built a glider, and promptly crashed and died. The Wright brothers did better. Wilbur Wright applied some knowledge about bird flight—that their wings warp as they maneuver—and experimented with a bendable box until he understood how to warp a wing's surface like birds do. By applying this analogy to wing design, Wilbur enabled his brother Orville to pilot an aircraft and survive the experience.

Analogical reasoning and Duncan's multiple-demand system seem to depend on cortical hub areas, but not the one in the temporal lobe. According to Duncan's idea, parietal and prefrontal areas draw on specialized representations in other parts of the cortex for various cognitive activities. Like the hub in the temporal lobe, which extracts the generalizations of cultural knowledge, parts of both the parietal and prefrontal cortex extract generalizations about relations in the world and successful solutions to problems. An improved ability to bring old solutions to bear on new problems provided your hominin ancestors with significant evolutionary advantages. All animals solve problems, of course, but humans do so at a level of generalization that no other species can match.

The areas that support Duncan's "multiple-demand system" have homologs in monkeys,[32] but that doesn't mean that they perform the same function or have the same kinds of representations. In Chapter 1,† we explained that bird wings and the

---

* To keep this chapter within bounds, we've given short-shrift to this crucial aspect of human cognition. John Duncan explains his ideas in detail in *How Intelligence Happens* (Yale University Press, New Haven, CT, 2012).
† In the section entitled "Homology."

front pair of fins in fishes are homologs. Yet, no one would claim that goldfish can fly simply because they have a homolog of bird wings. Although monkeys have prefrontal and parietal areas that are homologous to those in humans, we think that they took on new functions during hominin evolution. As the parietal and prefrontal regions expanded, they came to perform functions at a higher level of generalization, which depended on new kinds of neural representations. Recent research supports the idea that the most-expanded cortical areas in humans have extensive interconnections among each other and support a wide range of cognitive tasks.[33] By adding areas and neurons, new representations could emerge without the loss or degradation of older ones, so people got really good at solving puzzles like the one in Figure 9.5. (The correct choice is located in the lower left corner of the bottom array.)

Hominin generalizations reconsidered

In Chapters 7 and 8, we discussed several new representations that developed in anthropoids, each of which provided our ancestors with a specific evolutionary advantage: improved quantitative and qualitative vision; faster learning based on the memory of discrete goal-related events; and—as a result of all this—fewer foraging errors and a reduced risk of predation. In addition, although we haven't discussed it in any detail, anthropoids also evolved new representations of the sounds made by their own and other species.

The improvements in quantitative and qualitative vision were especially important. Anthropoids benefited from innovative representations of quantitative information in their parietal cortex—numbers, distances, and order, for example. They also possessed upgraded representations of qualitative information in their temporal cortex, including sights that served as signs of food and other resources. Because of their enhancements in vision, anthropoids—the descendants of nocturnal animals*—could make more effective use of vision while foraging in daylight.

The hominin innovations that we've discussed in this chapter depended on—and derived from—these anthropoid adaptations. As cortical hubs emerged in the expanding neocortex of evolving hominins, they added generalized hominin functions to the specialized ones they inherited from their anthropoid ancestors. Specifically:

- Part of the temporal lobe expanded and added a generalization—representing the concepts and categories of cultural memory—to a specialization: representing distant signs of food.

---

* As we explained in Chapter 6, early primates foraged at night.

- At about the same time, parietal areas also expanded and added a different form of generalization—representing relations of many kinds[34]—to a specialization: representing relations among quantities.

Both the temporal and parietal lobes interact with the prefrontal cortex, especially the parts that first emerged in anthropoids. Together, these evolutionary innovations improved the ability of evolving hominins to cope with novel and challenging problems. In this sense, hominin evolution produced cortical areas with a specialization for generalization.

We began this chapter by quoting *The Hitchhiker's Guide to the Galaxy* by Douglas Adams. In that story, a planet called Golgafrincham is home to an annoying array of administrators, middle managers, consultants, and telephone sanitizers. During a period of social tension, the thinkers and doers on Golgafrincham trick these ineffectual meddlers into traveling to Earth on a spaceship called the B-ark. The denizens of the B-ark expect that the A-ark, with the thinkers, and the C-ark, with the doers, will follow—but, of course, they never do. Regardless, the B-ark makes its way to Earth, and its passengers establish a new society, wiping out the indigenous hominin species in the process. (Sadly, back on Golgafrincham, the lack of telephone sanitizers leads to a plague that wipes out its civilization.) Readers eventually learn that human society functions as a computer, designed by mice to run a 10-million-year-long program to pursue the Great Question of Life, the Universe, and Everything—the answer to which is "forty-two." As luck would have it, Vogons destroy the Earth a few minutes before the program completes its computations. In an attempt to recover the results, a couple of mice study Arthur Dent—a descendant of the B-ark's passengers—in the hope that the Great Question resides somewhere in his brain. If it's in there, the mice would be well advised to begin their search among the newly evolved representations in his temporal lobe.

In the next and final chapter, we'll examine some consequences of the fact that human societies arose through evolution, rather than via "intelligent design" by mice (or anything else, for that matter). Along with cortical representations that enhanced generalizations, which we've discussed in this chapter, additional innovations arose in hominin brains. At first, these novel representations fostered cohesion and cooperation within hominin societies, but they eventually constructed a crucial part of the evolutionary road to human memory.

References

1. Andrews, P. *An Ape's View of Human Evolution* (Cambridge, UK: Cambridge University Press, 2015).
2. Hill, J. et al. Similar patterns of cortical expansion during human development and evolution. *Proceedings of the National Academy of Sciences U.S.A.* **107**, 13135–13140 (2010).

3. Sekeres, M. J. et al. Changes in patterns of neural activity underlie a time-dependent transformation of memory in rats and humans. *Hippocampus* **28**, 745–764 (2018).

4. Essig, G. *Lesser Beasts: A Snout-to-Tail History of the Humble Pig* (New York, NY: Basic Books, 2015).

5. Binder, J. R. & Desai, R. H. The neurobiology of semantic memory. *Trends in Cognitive Sciences* **15**, 527–536 (2011).

6. Binder, J. R., Desai, R. H., Graves, W. W., & Conant, L. L. Where is the semantic system? A critical review and meta-analysis of 120 functional neuroimaging studies. *Cerebral Cortex* **19**, 2767–2796 (2009).

7. Glasser, M. F., Goyal, M. S., Preuss, T. M., Raichle, M. E., & Van Essen, D. C. Trends and properties of human cerebral cortex: correlations with cortical myelin content. *Neuroimage* **93P2**, 165–175 (2014).

8. Van Essen, D. C. & Dierker, D. L. Surface-based and probabilistic atlases of primate cerebral cortex. *Neuron* **56**, 209–225 (2007).

9. Visser, M., Jefferies, E., Embleton, K. V., & Lambon Ralph, M. A. Both the middle temporal gyrus and the ventral anterior temporal area are crucial for multimodal semantic processing: distortion-corrected fMRI evidence for a double gradient of information convergence in the temporal lobes. *Journal of Cognitive Neuroscience* **24**, 1766–1778 (2012).

10. Hodges, J. R., Graham, N., & Patterson, K. Charting the progression in semantic dementia: implications for the organisation of semantic memory. *Memory* **3**, 463–495 (1995).

11. Cappelletti, M., Butterworth, B., & Kopelman, M. Numeracy skills in patients with degenerative disorders and focal brain lesions: a neuropsychological investigation. *Neuropsychology* **26**, 1–19 (2012).

12. Bozeat, S., Lambon Ralph, M. A., Patterson, K., & Hodges, J. R. When objects lose their meaning: what happens to their use? *Cognitive, Affective, and Behavioral Neuroscience* **2**, 236–251 (2002).

13. Bozeat, S., Lambon Ralph, M. A., Patterson, K., Garrard, P., & Hodges, J. R. Non-verbal semantic impairment in semantic dementia. *Neuropsychologia* **38**, 1207–1215 (2000).

14. Rogers, T. T., Patterson, K., & Graham, K. S. Colour knowledge in semantic dementia: it is not all black and white. *Neuropsychologia* **45**, 3285–3298 (2007).

15. Rogers, T. T., Lambon Ralph, M. A., Hodges, J. R., & Patterson, K. Natural selection: the impact of semantic impairment on lexical and object decision. *Cognitive Neuropsychology* **21**, 331–352 (2004).

16. Bozeat, S. et al. A duck with four legs: investigating the structure of conceptual knowledge using picture drawing in semantic dementia. *Cognitive Neuropsychology* **20**, 27–47 (2003).

17. Acosta-Cabronero, J. et al. Atrophy, hypometabolism and white matter abnormalities in semantic dementia tell a coherent story. *Brain* **134**, 2025–2035 (2011).

18. Visser, M. & Lambon Ralph, M. A. Differential contributions of bilateral ventral anterior temporal lobe and left anterior superior temporal gyrus to semantic processes. *Journal of Cognitive Neuroscience* **23**, 3121–3131 (2011).

19. Spitsyna, G., Warren, J. E., Scott, S. K., Turkheimer, F. E., & Wise, R. J. Converging language streams in the human temporal lobe. *Journal of Neuroscience* **26**, 7328–7336 (2006).

20. Lambon Ralph, M. A. Neurocognitive insights on conceptual knowledge and its breakdown. *Philosophical Transactions of the Royal Society London B Biological Science* **369**, 20120392 (2014).

21. Rosch, E., Mervis, C. B., Gray, W., Johnson, D., & Boyes-Braem, P. Basic objects in natural categories. *Cognitive Psychology* **8**, 382–439 (1976).
22. Rogers, T. T. & Patterson, K. Object categorization: reversals and explanations of the basic-level advantage. *Journal of Experimental Psychology, General Psychology* **136**, 451–469 (2007).
23. Lambon Ralph, M. A., Lowe, C., & Rogers, T. T. Neural basis of category-specific semantic deficits for living things: evidence from semantic dementia, HSVE and a neural network model. *Brain* **130**, 1127–1137 (2007).
24. Lambon Ralph, M. A., Ehsan, S., Baker, G. A., & Rogers, T. T. Semantic memory is impaired in patients with unilateral anterior temporal lobe resection for temporal lobe epilepsy. *Brain* **135**, 242–258 (2012).
25. Schapiro, A. C., McClelland, J. L., Welbourne, S. R., Rogers, T. T., & Lambon Ralph, M. A. Why bilateral damage is worse than unilateral damage to the brain. *Journal of Cognitive Neuroscience* **25**, 2107–2123 (2013).
26. Lambon Ralph, M. A., Pobric, G., & Jefferies, E. Conceptual knowledge is underpinned by the temporal pole bilaterally: convergent evidence from rTMS. *Cerebral Cortex* **19**, 832–838 (2009).
27. Pobric, G., Jefferies, E., & Lambon Ralph, M. A. Anterior temporal lobes mediate semantic representation: mimicking semantic dementia by using rTMS in normal participants. *Proceedings of the National Academy of Sciences U.S.A.* **104**, 20137–20141 (2007).
28. Crone, E. A. et al. Neurocognitive development of relational reasoning. *Developmental Science* **12**, 55–66 (2009).
29. Mackey, A. P., Miller Singley, A. T., & Bunge, S. A. Intensive reasoning training alters patterns of brain connectivity at rest. *Journal of Neuroscience* **33**, 4796–4803 (2013).
30. Duncan, J. The multiple-demand (MD) system of the primate brain: mental programs for intelligent behaviour. *Trends in Cognitive Sciences* **14**, 172–179 (2010).
31. Woolgar, A. et al. Fluid intelligence loss linked to restricted regions of damage within frontal and parietal cortex. *Proceedings of the National Academy of Sciences U.S.A.* **107**, 14899–14902 (2010).
32. Mitchell, D. J. et al. A putative multiple-demand system in the macaque brain. *J Neurosci* **36**, 8574–8585 (2016).
33. Sneve, M. H. et al. High-expanding regions in primate cortical brain evolution support supramodal cognitive flexibility. *Cerebral Cortex*, doi:10.1093/cercor/bhy268 (2018).
34. Genovesio, A., Wise, S. P., & Passingham, R. E. Prefrontal-parietal function: from foraging to foresight. *Trends in Cognitive Sciences* **18**, 72–81 (2014).

10

The story of your life

Memories all your own

PROFESSOR MARVEL [GAZING INTO HIS CRYSTAL BALL]: Let's see—you're . . . traveling in disguise. No, that's not right . . . You're—running away.

DOROTHY: How did you guess?

PROFESSOR: Professor Marvel never guesses—he knows! Now, why are you running away? . . . No, no—now don't tell me. They . . . don't understand you at home. They don't appreciate you. You want to see other lands—big cities—big mountains—big oceans.

DOROTHY: Why, it's just like you could read what was inside of me.

As indeed he could. One part of this chapter will deal with the wizardry of mind reading, another with the origin of personal memory. First, though, we'll take up a few points about human evolution.

Explosions

CHAPs and change

In Chapter 9,[*] we mentioned the population of apes that gave rise to both humans and chimpanzees. Many people assume that this species, which we call the CHAPs (the *chimpanzee–human ancestral population*), closely resembled modern chimps. But they differed in important ways. For example, the CHAPs had hands that were more like ours and less like those of modern chimpanzees, which have longer fingers and shorter thumbs. We've inherited our relatively short fingers from the CHAPs, and the long digits of chimps evolved in their separate branch of the family tree. As a result, chimpanzee hands have less manipulative capability than either human hands or those of the CHAPs. Although the CHAPs used and made tools, there's no evidence that they used tools to manufacture other tools, as some of our hominin ancestors did. Females and males also had a bigger size difference than in modern humans. Modern men are about 15% larger than women, on average; for the CHAPs, adult males were about 50% larger.[1]

[*] In the section entitled "CHAPs, not chimps"

The Evolutionary Road to Human Memory. Elisabeth A. Murray, Steven P. Wise, Mary K. L. Baldwin, and Kim S. Graham, Oxford University Press (2020). © Oxford University Press.
DOI: 10.1093/oso/9780198828051.001.0001

In the hominin branch of the family tree, most species resembled the CHAPs at first, but one of them soon adopted a more upright form of locomotion as they spent more time at ground level and less in trees. Early hominins also developed smaller canine teeth, probably because they ate the tougher foods found at ground level. A diet rich in hard foods—such as nuts, seeds, and tough roots—led to an increase in the size of premolars and molars, along with a decrease in the size of teeth toward the front of the mouth. All of this produced the relatively flat faces that characterize hominins.[1]

Later hominins came to differ from the CHAPs in additional ways. Sometime after they abandoned the shelter provided by tropical woodlands for more open terrain, hominins developed stone technology. Simple manufactured stone tools appeared 2–3 million years ago, including tools used to make other tools. Then, some 1–2 million years ago, a varied and sophisticated toolkit appeared, including two-sided stone axes. The ability to control fire also emerged at about this time, and some social changes were likewise afoot. In these ancestral species, gender-based differences in size diminished to the 15% level typical of modern humans, which suggests that the hominins of this time had adopted a new social system. According to Richard Klein,[2] an expert on human evolution, more equal size could "mark the beginnings of a distinctively human pattern of sharing and cooperation between the sexes, prefiguring the social organization of historic hunter–gatherers."

Brains, burns, and bangs

Hominins began with brains barely larger than those of modern chimpanzees (relative to body weight). Then, over the past 3 million years,[2] the hominin brain expanded until it reached its current size somewhere around 400 000 years ago.[3] In the interim, the toolkits of hominins changed in concert with alterations in the anatomy of their brains and bodies. After human brains reached their current size, the rate of cultural innovation accelerated but human anatomy changed relatively little.[2] Eventually, these people manufactured musical instruments, drew animals on the walls of caves, carved figurines in human and animal form, made ornaments for their bodies, buried their dead in elaborate graves, and fashioned bones and seashells into pins and needles, among other cultural innovations.

Experts in human evolution regularly (and vigorously) debate this brief summary, and many uncertainties remain. Some authorities believe that the culture just described developed gradually over a few hundred thousand years; others think that this culture emerged abruptly as recently as 60 000 years ago. Proponents of the latter view sometimes say that a "creative explosion" occurred at this time, after which the rate of cultural innovation accelerated rapidly. These are fascinating debates, but for our purposes they don't matter much. A few hundred thousand years doesn't differ much from 60 000 years on the timescale of primate evolution. Whether human cultural innovations accelerated within the past 60 000 years or four times that long, it amounts to only a small fraction of 1% of primate history. So, whether human culture materialized from a big bang or a slow burn, the term "creative explosion" captures what happened well enough for our purposes. Because this phrase is a bit bombastic for some scientists, we place it in quotation marks.

Before the blast

From this timeline, it's clear that the human brain reached its current size substantially before an abrupt spurt of inventiveness, if there was one, or a little before a more gradual acceleration if that's what happened. What's more, the "creative explosion" occurred long after hominins shifted to walking upright, adopted a new diet, devised an elaborate tool-making culture, and developed a social system characterized by decreased differences between the sexes in size and strength.

There's also evidence, albeit indirect and far from conclusive, that human language originated long before any "creative explosion," maybe sometime around 400 000–800 000 years ago.[4] A human gene, called *FoxP2*, has attracted a lot of attention, in part because of its romantic designation as "the language gene." This gene adopted its modern form somewhere around 400 000–800 000 years ago in a human species that gave rise both to modern humans and Neanderthals.[5] This finding suggests to some scientists that Neanderthals had language, an idea supported by the anatomy of a bone that supports muscles in their voice box and tongue.[6,7] So, an origin of language 400 000–800 000 years ago seems reasonable. Research has shown, however, that the so-called "language gene" isn't specific to language at all; its influence extends to all kinds of coordinated movements of the mouth, lips, tongue, and face.[8] Fortunately, other evidence also supports these dates for the origin of symbolic language—or at least a proto-language of the sort we'll mention later. Robin Dunbar[4] has summarized the evidence as follows: about 800 000 years ago the vocal tract underwent modifications that would have altered the sounds that hominins could make. At about the same time, passageways through certain bones widened, which permitted a greater number of nerve fibers to pass through them on the way to muscles of the mouth and tongue. This development would have enhanced fine control over muscles used for speech. And about 500 000 years ago, the ear canal changed in shape, which probably enhanced the detection of high-pitched sounds. Sounds in this tonal range play a major role in understanding speech. So, although the jury remains out, it seems likely that speech and language arose long before the "creative explosion," too. Symbolic language certainly facilitated an acceleration in cultural innovation, but if Dunbar's estimate for the origin of language is anywhere close to accurate, something additional needed to happen.

Innovation and identity

If neither upright locomotion, a new diet, the manufacture of toolkits, symbolic language, nor social changes lit the fuse for a "creative explosion," what did? At the moment, expert opinion offers two general answers: one emphasizes changes in the brain and cognition; the other stresses social factors.[9] Our proposal combines these ideas and adds some anatomical specificity. We think that a "creative explosion" resulted from two evolutionary developments: (1) changes in the brain that led to what we'll call *enhanced generalization*; and (2) a new kind of representation of the "self" adapted to hominin social systems—although not necessarily in that order.

- We discussed enhanced generalization in Chapter 9. There we described a cortical hub in the temporal lobe that extracts the generalizations of cultural (semantic) memory, including concepts about the world and categories of objects, plants, animals, and people in it. Similar evolutionary developments enabled parietal areas to extract generalizations about relations, which underlie the ability to bring old solutions to bear on new problems, partly through analogical, metaphorical, and relational reasoning.
- We'll take up representations of the "self" later in this chapter. There, our discussion will revolve around a cortical hub that evolved in the prefrontal cortex and the idea that it extracts generalizations about the "self" from representations of an individual's intentions and actions.[10]

Bigger is better

These evolutionary developments went hand-in-hand with an expansion of the frontal, temporal, and parietal lobes, which we illustrate in Figure 9.1a. The cerebral cortex makes up about 80% of the hominin brain by volume, and recent research has revealed that as the brain expanded, the number of cortical neurons increased compared to other primates. The last common ancestor of Old World monkeys and humans probably had about 1.5 billion neurons in its cerebral cortex; you have about 16 billion in yours[11]: an approximately 10-fold increase that comes along with a massively increased matrix of interconnections among these neurons. Earlier, we emphasized the emergence of new cortical areas during evolution, with which came novel kinds of neural representations. But new representations can emerge without new areas, either from an increase in the number of neurons and their interconnections within an area or through new interactions among cortical areas.

The idea that the prefrontal cortex expanded dramatically during hominin evolution is important for our proposal—and somewhat controversial. It shouldn't be.

In the left part of Figure 9.1a, a big yellow blob near the front of the brain indicates that the prefrontal cortex expanded as much as any region—and more than most.[3] The reason for controversy is that some experts say that the human brain doesn't have any more prefrontal cortex than expected for a primate with a cerebral cortex of its size.[12] "Expected" is a key word here because what that often means is that as the cerebral cortex expanded during hominin evolution, the prefrontal cortex simply kept pace with this overall expansion. But even if this is so, the human prefrontal cortex is still 4–5 times larger than in chimpanzees (and other apes) in terms of its absolute volume, and there is good reason to believe that the hominin prefrontal cortex expanded much more than the remainder of the frontal lobe.[3] (The frontal lobe includes the prefrontal, premotor, and primary motor areas.) Because of this expansion, the prefrontal cortex makes up about 80% of the human frontal lobe, compared to 55% or so in chimpanzees and 40–50% in monkeys.[13] By one analysis, humans have about three times more prefrontal cortex in the frontal lobe than expected for an anthropoid brain.[3] The question is: what does all this extra prefrontal cortex do? And what did it do when it first evolved?

Me, myself, and I

One answer is that it contributed to novel representations of the "self," which provided evolving hominins with advantages in terms of social cohesion and co-operation. Evidence for this idea comes mainly from the anatomical pattern of increased cortical activations observed during brain-imaging experiments. These studies point to the prefrontal cortex as a cortical hub for representations of the "self." Most of the highly activated areas lie in middle parts of the frontal cortex,* where the two hemispheres face each other. (In Fig. 1.2b, we illustrate these regions by including only one of the two hemispheres.)

In brain-imaging experiments, parts of the prefrontal cortex show an increased level of activation when people contemplate themselves or others. Significant increases in activation occur when people pay attention to their own actions, intentions, or internal states such as hunger, emotions, or heart rate.[10] Something similar happens when people evaluate whether a particular trait applies to themselves; when they evaluate their own behavior or personality traits; when they remember an event involving themselves; when they consider the traits of other people or make social judgments about them; when they receive evaluations about themselves from others; and when they consider how to improve someone else's emotional state.[14-19]

---

* Scientists often refer to these areas by names such as the medial frontal cortex, medial prefrontal cortex, and medial frontal pole cortex.

Insight into others

Many of the same or nearby areas also represent "others,"[15] meaning other people. The ability to represent the "self" and your own mental states apparently evolved in harmony with an ability to represent the mental states of other people. We don't know which came first, but in either case the result was a cognitive capacity called a *theory of mind*, which enables people to infer the perceptions, knowledge, intentions, and internal states of other people.[19, 20] Dorothy alluded to this capacity in the epigraph of this chapter; as she said to Professor Marvel: "it's just like you could read what was inside of me."

In some cases, your penchant for mind reading can cause comical illusions. When circles move around on a video display and collide with other circles, deflect them, and so forth, you'll perceive each circle as having intentions and emotions.[21] These illusions are almost impossible to resist despite the fact that you know that circles don't have feelings and have no intention of doing anything.* When it first evolved, the automatic nature of these perceptions probably promoted social cooperation, but all sorts of unrelated consequences flowed from this development. Animism, for example, stems from an innate tendency to attribute intentions and emotions to inanimate objects or natural phenomena. Hurricane Irma, with its cyclopean cyclonic eye, is said to have "taken aim" at Florida, just before "lashing the coast" with "fury." Or so it seems.

The "self," society, and the in-body experience

Any animal that remembers its own actions represents itself to some extent. So, representations of the "self" and "others" are probably widespread among animal species and especially among our closest relatives. The species-specific representations that evolved in hominins provided advantages for their social systems, including recognizing intentions, sharing knowledge, and following social conventions.[4] Similar representations developed in other species in support of their social systems.

Mike Graziano and Sabine Kastner[15] have suggested that high-level representations of the "self" result in part from attributing perceptions and intentions to a person's own body. They cite, for example, an experiment that used disruptive electromagnetic stimulation of the brain to induce an out-of-body experience.[22] The cortical area manipulated in this experiment has, in other studies, shown increased activation in tasks involving personal agency,[23] a term that refers knowledge about how one's own actions affect events. From these observations, it seems likely that the ability to localize thoughts and feelings to one's own body—the

---

* One example, from *Sesame Street*, can be found at: https://www.youtube.com/watch?v=i1-z9rJYHoc

"in-body experience"—depends on sophisticated representations of the "self," including what the "self" achieves, plans, and perceives.

Insight as the "self"

When you attend to your brain's representations of your own intentions, you perceive them in much the same way as you perceive sensations. This idea seems strange at first, but all perception depends on attending to representations in the brain, usually in "sensory areas" of cortex. You have the subjective impression that when you perceive something—a dim green light across a foggy lake, for example—you're attending to something in the outside world. But representations in the brain always come into play. In this example, the relevant representations are located in visual areas of the cortex. Your perception of a green light depends as much on representations in the brain as when you imagine a green light with your eyes closed. So, even your simplest perceptions depend on insight, in the literal sense of "seeing" inside your own brain. In *The Great Gatsby*, the narrator notices Gatsby tremble as he gazes at a dim green light on a distant dock. As he does, Gatsby not only attends to sensory representations in his visual cortex, but also to his intentions toward Daisy Buchanan, his long-lost love.

There is, however, an important anatomical difference between attention to sensations and attention to intentions. The former depends on interactions between the frontal lobe and "sensory areas" toward the back of the brain; the latter depends on interactions mainly within the frontal lobe. According to this idea, as the prefrontal cortex expanded during hominin evolution it added new and higher levels of representation, including representations of the "self."[24] The expanded prefrontal cortex established these representations by extracting information from "lower-ranking" representations of intentions and actions elsewhere in the frontal lobe. This idea resembles the one we advanced in Chapter 9 for the temporal lobe. There we said that higher-level representations in a temporal hub area resulted from extracting information for lower-level representations in specialized spoke areas. The same idea goes for a hub in the prefrontal cortex.

We don't mean to imply that the "self" and "others" are embodied wholly within the prefrontal cortex. These representations probably originate in the prefrontal cortex, but once generated they become available to other brain areas—in abundant combinations. Two large-scale cortical networks, which we call *participatory* and *cultural*, distribute these representations to a large portion of the brain.

The participatory network

By the *participatory network*, we refer to an extended set of interconnected cortical areas that underlie a sense of participation in events, both as they occur and when recalled from memory.* A sense of participation is crucial to human memory, and it has many consequences, including the construction of a mental autobiography.

The participatory network includes many of the areas that we illustrate in the top part of Figure 1.2b, along with the hippocampus, which appears in Figure 1.2d. As we explained in Chapter 4, the hippocampus evolved in early vertebrates to support navigational representations. As mammals evolved, several new cortical areas emerged and began to interact with the hippocampus, including a group of nearby areas.† By embedding the "self" within their various representations, new and old cortical areas work together to establish a sense of participating in events. The exchange of information among components of the participatory network, including the medial prefrontal cortex, means that no single area is responsible for participatory memories or a mental autobiography.

The participatory network provides much more than narcissistic fascination; it also empowers the construction of imaginary events. From this capacity comes every novel ever written, every story ever told, and—as probably its most important consequence—a sense of participation in an imaginary future in which the "self" is the star of the show. Through the mental simulation of events, people can attempt to understand the past and how things might have turned out differently. Similar mental simulations allow people to imagine various strategies for solving a novel problem and what might happen if they implement a potential solution. The advantage of such mentation can be, if all goes well, avoiding mistakes rather than having to learn from them. Combined with a theory of mind (and, specifically, its inferences about the mental states of other people), mental simulations produce a certain kind of magic; in *The Wizard of Oz*, Professor Marvel not only knows Dorothy's desire to see new places and big cities, but also that she would head home when fed a fib about Aunt Em: "someone has just about broken her heart," he lies.

In addition to terms like *mental simulation*, related labels for the participatory network's functions include future-thinking, mental trial-and-error, foresight, prospection, future modeling, scenario construction, and constructive episodic simulation. By any name, the participatory network represents both real experiences and imaginary ones, including revisionist and counterfactual histories. The same kinds of representations that underlie the memories of events experienced in real life also do so for fictional events. Accordingly, some parts of the participatory network exhibit a high level of activation when people recall fictional memories; others do so for factual memories.[25]

All of these memories come with a sense of personal participation: not only for the events of an individual's life, but also for events that happened to other people and to fictional characters. The same goes for events that occur in the absence of

---

* Scientists often refer to this interconnected set of cortical areas as the default-mode network or the medial network.

† Recall that the hippocampus is an allocortical area, in contrast to neocortex. The nearby areas that interact with it include the retrosplenial cortex, the posterior cingulate cortex, the entorhinal cortex, the subicular cortex, and the parahippocampal cortex, among others.

anyone. Opinions differ about whether a falling tree makes a sound when no one's around to hear it, but no one doubts the ability to imagine a tree falling in an unoccupied forest. So, to understand what we're suggesting, you need to construe the word *participatory* broadly. Some experts refer to the fruits of the participatory network as *mental time travel*, a term that captures the sense of experiencing (or "reliving") both remembered and imagined events. Mental time travel can include elaborate simulations of events that never occurred—and, in many instances, never could. As Mark Twain once wrote: "I've lived through some terrible things in my life, some of which actually happened." People have no difficulty imagining a dim green light across a foggy lake, even if they've never seen one; and just about anybody can identify with Jay Gatsby as he daydreams about a future with Daisy Buchanan. Through identification with fictional characters, people can experience a sense of participation in something wholly fabricated, a story that does not involve themselves in any way. It's *participatory* all the same.

The evolutionary advantages of the participatory network are obvious: it enabled evolving hominins to imagine what other members of their social group had in mind, as well as to establish the common ground and shared intentions required for social cooperation. It also promoted an appreciation of future prospects, along with the ability to adopt multiple perspectives for considering a given situation. All of these capacities promoted survival in the close-knit social groups of ancestral hominins.

The cultural network

The name *cultural network* draws on our discussion in Chapter 9. There we described representations that underlie the generalizations, concepts, and categories of cultural knowledge. For this chapter, a key point about cultural knowledge is that it includes social knowledge, including concepts about individuals and groups and their contributions to society. Social concepts include knowledge about personality types (brooders and blowhards), occupations (athletes and actors), kinship categories (nieces and nephews), maturity states (toddlers and tweeners), body types (strong and scrawny), status (leaders and losers), and inquisitiveness (science enthusiasts and science deniers), among others. All of these representations can be associated with a range of emotions and valuations; after all, no one wants to be a scrawny, science-denying blowhard of a loser.

The cultural network includes the cortical hub that we discussed in Chapter 9, which occupies the front part of the temporal lobe. It also encompasses the hub's specialized spoke areas, which we illustrate in Figure 9.3. The hub extracts generalizations from spoke areas about groups of people and their roles in society,[26] just as it does for objects, plants, and animals. The cultural network encompasses all of this information. Spoke areas include parts of the temporal lobe, the prefrontal cortex, and a region where the parietal and temporal lobes merge into one another, among other cortical areas. Some spoke areas specialize in representing the faces of individuals and emotional expressions; others represent body language, which conveys social signals such as "get lost loser."

Interactions between the participatory and cultural networks are particularly important, and here we consider just one. Language is, at heart, a social behavior aimed at influencing or informing other people through a mutual understanding of word meaning and grammar. Although we don't commonly say it this way: speakers seek to influence neural representations in the brains of other people. The linguist James Hurford[27] has suggested that the earliest languages had what he calls a topic–comment grammar. In his proposed proto-language, the first word of a two-word sentence directs a listener's attention to something. The rest of the sentence—its second word—consists of a comment about that topic. A topic–comment grammar draws on both the participatory and cultural networks. The participatory network provides the basis for understanding that other individuals have cognitive representations to influence; the cultural network provides the comment's content. "Lion hungry"—a topic and a comment—would send an important message to a fellow hominin who had to live among both dangerous, hungry lions and docile, sated ones.

The "self" and society

Innovative representations of the "self" and "others" endowed ancestral hominins with a form of wizardry: an ability to recognize themselves in others and others in themselves. By imagining the intentions of other people—either in the immediate future or in some distant one—our ancestors gained important advantages in fashioning human societies: cooperation in achieving shared goals; cultural memories involving kinship, status, and personality types; and understanding the contributions that an individual can (or should) make to a social group.

The birth of personal memory

Once new and higher levels of "self" representation evolved in an expanding prefrontal cortex, they became available to both the participatory and cultural networks. Together, these two networks and their interactions created a powerful capacity for incorporating species-specific representations of the "self" into memories. We illustrate this idea in Figure 10.1: embedding the "self" into the memory of an event establishes a participatory (episodic) memory; and embedding the "self" into the memory of a fact or a fiction establishes a cultural (semantic) memory. The second idea is a little trickier than the first. At first glance, you might think that you don't attach your "self" to your knowledge about *The Wizard of Oz*. And it's true that you don't think of your "self" as part of your knowledge that the ruby slippers have magical powers. Instead, think of the "self" as a secret, silent attachment to your knowledge about ruby slippers, which when integrated into that memory conveys a sense that the memory is yours: that you own it. Much the same goes for participatory memories. You don't think that you actually participated in melting the Wicked Witch of the West, but the silent "self" integrated into your memory of this fictional event enables you to "relive" it in your imagination. We think that when the composite representations depicted in Figure 10.1 arose in hominin evolution, personal memories—the sense of participating in events and owning knowledge—came into the world for the first time.

Take, for example, the kinds of representations that we discussed in Chapter 4. As we explained there, early vertebrates evolved a homolog of the hippocampus,

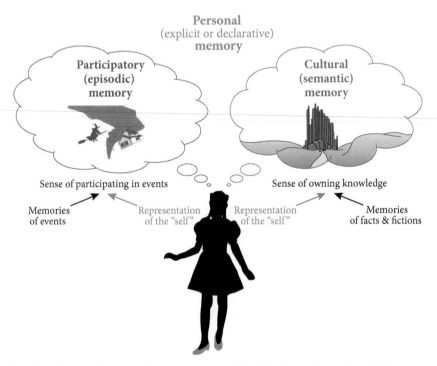

Fig. 10.1 Personal memory. Representations of the "self" combine with both the memories of events and the memories of facts and fictions. These combinations produce a sense of participation in events and ownership of knowledge. In *The Wizard of Oz*, Dorothy can remember what it was like to be in a tornado-propelled house, illustrated to the right of the funnel cloud, as well as the experience of seeing the Wicked Witch of the West flying on her broomstick. Likewise, she remembers a thing or two about the Emerald City, such as its location at the end of the yellow brick road.

which supported novel neural representations that guided navigation through a complex seascape. We also described how the same kinds of representations record events, sequences, scenes, situations, and perspectives more generally. In modern humans, the participatory network supports these representations, with the hippocampus playing a prominent role. When neural representations of your "self" become attached to these memories, they accumulate as a mental autobiography.

By thinking about human memory in this way, we've concluded that animals neither remember a personal autobiography nor have a sense of participating in events or owning knowledge.* In other words, animals don't have personal (explicit) memories. The reason is that they don't have representations of the "self" that are exactly like the ones that evolved in humans. And they lack these representations because they don't have the expanded prefrontal areas that generate and store them. As simple and sensible as this idea is, you should know that there isn't yet any experimental evidence to back it up, which is why few neuroscientists accept it for the time being. For now, let's just say that it makes sense to us. Most scientists will continue to assume that animals have the same kinds of memories as people do, in part because they use terms like episodic and semantic memory in ways that don't necessarily imply the sense of participation in events or ownership of knowledge that (we think) makes human memory unique.

The prospect of having knowledge without a sense of owning that knowledge might seem weird, but your brain contains many hidden memories. For example, like many anthropoids, you're predisposed to look at faces. Recent evidence indicates that the amygdala contains instinctive knowledge that guides anthropoids like you to scrutinize faces with the fovea.[28] In Chapter 6, we mentioned several additional kinds of memories that lurk "unowned" in your brain: how to detach an apple from a tree, for example. In the tradition of Isaac Newton, you also know that when an apple naturally detaches from its stem, it'll fall toward the ground. Many grammatical rules reside in your brain without the sense that you own them— or even *have* them. You pick up these rules automatically, mostly during childhood. People who don't know whether a *gerund* is a grammatical term or a small, furry animal can still use verbs as nouns, and that's the definition of a gerund. In Chapter 1, we referred in a footnote to a former American president dissembling about what "the meaning of the word is is." The first "is" in the phrase "is is" is a gerund.

Not only do animals lack personal memories, they also experience emotions differently than people do. In Chapter 3,† we mentioned the fact that animals respond to threatening sensations with defensive actions and reactions. Examples of the former include active escape or avoidance; the latter include freezing in

* In *The Gap: The Science of What Separates Us from Other Animals* (Basic Books, New York, NY, 2013), Thomas Suddendorf presents a thorough analysis of how human and animal cognition differ.
† In the section entitled "Phony parsimony."

place and autonomic responses. These behaviors don't mean that animals have the human experience of fear or anxiety. Without the species-specific representations of the "self" that our species has, animals probably don't experience emotions and moods as you do. Two experts on emotion, Joe LeDoux* and Danny Pine, have explained these ideas extraordinarily well.[29] Independently, they set out to understand human emotion at about the time that we set out to explore human memory. Remarkably, the two inquiries led to similar conclusions, probably because the subjective experience of emotion and the sense of participation in events both depend on the same hominin-specific representations of the "self."

Once high-level, species-specific representations of the "self" evolved and gained wide cortical distribution via the participatory and cultural networks, personal memories provided hominin societies with several advantages. As we mentioned earlier, our ancestors could consider knowledge about their social group, including the intentions and feelings of other people. Consequently, they could use shared intentions and common knowledge in the cooperative pursuit of goals. In addition, individuals could re-experience past events and reflect upon them in the context of an imaginary future.

The cross-domain cathedral

Even if a hominin species had personal memories—including representations of the "self" and a sense of participation in life—that doesn't mean that they had the kind of mind that ignited the "creative explosion." A modern human mind, and the "creative explosion" that it fostered, required something more, and we think that this "something more" might have been the evolutionary changes summarized in Chapter 9. There we explained that an expansion of the temporal, parietal, and prefrontal cortex accompanied the emergence of new and higher levels of representation, which supported two innovative forms of generalization: one for cultural memories; the other for relations. We call these enhanced capacities *enhanced generalization*, which provided evolving hominins with new and more-general levels of knowledge about both the world and themselves.

Stephen Mithen[30] has developed a useful metaphor for how enhanced generalization might have evolved. Before the "creative explosion," he suggested, the hominin mind resembled a medieval cathedral that included several "chapels" walled off from each other. Each "chapel" corresponded to a particular kind of specialized knowledge: about animals and plants; about society; or about tools and other objects. All of these categories of knowledge, also known as *cognitive domains*, contributed to the success of our hominin ancestors. The "creative explosion" occurred

---

* In his book, *Anxious: Using the Brain to Understand and Treat Fear and Anxiety* (Penguin, New York, NY, 2015) Joe LeDoux explains these ideas in detail.

after some development breached the "walls" between the "chapels." After that, the rate of cultural innovation accelerated because different cognitive domains could interact with and enrich each other. For example, memories about animals could inform tool-making and *vice versa*. So, people could understand how to use animal bones, ivory, and shells to fashion awls, needles, pins, and other tools that were hard to make from stone. People could also use their social knowledge to predict what animals might do. After the "walls" were breached, every cognitive domain could contribute to solving problems that had previously depended on only one domain. So, cultural advances accrued very quickly.

To extend this metaphor, we can ask: what breached the "walls" of Mithen's "chapels"? According to our proposal, the evolutionary developments that we discussed in Chapter 9 might have served as the final breakthrough. Symbolic language certainly contributed to cultural advances, but even that development did not suffice to ignite the "creative explosion," which probably came much later.* Perhaps the key development involved the widespread interconnectedness that developed among expanded neocortical areas of the temporal, parietal, and frontal lobes. Scientists have recently confirmed that the most-expanded parts of the temporal, parietal, and frontal lobes have profuse connections with each other and show an enhanced level of activation when a task requires information from several different cognitive domains.[31] During human evolution, cortical expansion—as depicted in Figure 9.1a—enabled linkages among the hundreds of millions of additional neurons that populated these areas.[11] We suspect that these new connections broke down Mithen's "chapel walls" and fostered cross-domain knowledge, cross-domain reasoning, and cross-domain problem-solving. Afterward, ancestral humans could develop a potent storehouse of knowledge combined with species-specific representations of the "self," which probably evolved in earlier hominins. When all of these elements came together, personal (explicit) memory had become fully modern and uniquely human. At that point in evolution, early modern humans could exchange their personal memories through language, which enabled cultural advances to accumulate over generations.

What can possibly go wrong?

As happens often in evolution, developments that provide advantages also create vulnerabilities. When the red arrows in Figure 10.1 break down, serious problems can result. The ability to imagine events, for example, leads to a requirement for *reality monitoring*, which refers to the capacity for distinguishing actual events from imaginary ones. Sadly, some people lose this capacity,

---

* As we explained in the section entitled "Brains, burns, and bangs."

possibly because they can't attach a representation of their "self" to something they imagine. They also don't know whether an intention is their own, leading to a sense of control by external forces. As they mentally rehearse language, an inability to link these thoughts to representations of their "self" makes it seem as though voices come from somewhere or something else. These are some of the symptoms of schizophrenia, a mental disorder in which people sometimes don't realize that remembered sensations, thoughts, and intentions arise from their own brains.

Consciousness not explained

In *Consciousness Explained*, the philosopher Daniel Dennett[32] characterizes two kinds of false memories. He calls one Stalinesque, an allusion to Stalin's show trials of the 1930s. People remember what they've experienced, but what they've experienced was an illusion. In the Orwellian form of false memory—an homage to the rewriting of history in Orwell's novel *1984*—people revise their memories in light of their current knowledge and opinions.

We mention Dennett's ideas because of his book's title, which we've modified for this section's heading. Many readers, no doubt, will be disappointed that we've avoided consciousness and its neural correlates. For most scientists, discussing consciousness is like shearing a pig for its wool: the meager quantity of fleece isn't worth all the squealing. Scientists often contrast human consciousness with cognition in chimpanzees or other modern species, but our evolutionary perspective leads us to view the issue differently. When we consider the dozen or so extinct hominin species that stretch back more than five million years, it's clear that no one knows the crucial factors that distinguish modern human consciousness from earlier human varieties or from whatever yet-more-distant hominin ancestors might have experienced. The ability to appreciate the ownership of knowledge, for example, could be like ownership of material possessions: sometimes conscious and sometimes not.

A *priming* experiment demonstrated subconscious ownership. In priming, a subliminal sensation activates representations in the brain, which has an influence on perception and behavior. In this experiment, people first saw a series of pictures and were told that they owned each one. Later, they had to sort words, presented one at a time, into two categories: words about themselves (such as me, myself, and I) and words about others (such as they, them, and theirs). When people had seen one of their own pictures just before seeing a word, they sorted the words faster and more accurately than after they had seen a picture they didn't "own." Their "ownership" of the pictures influenced their behavior even though they weren't aware of it as they sorted the words.[33] If the ownership of pictures can be subconscious, so can the ownership of knowledge.

In a brain-imaging study,[14] which also involved priming, the experimenters localized the awareness of knowledge about the "self" to the participatory network. Like the previous experiment, people saw words on a video screen and sorted them into two categories: self-related and other-related. This experiment used *perceptual masking*, a procedure that could, if the timing was right, block awareness of a word. Nonsense letters appeared shortly after the word—on the same part of the screen. If a sufficiently long time elapsed between the self-related word and the nonsense letters, people became conscious of the word and its relationship to themselves. In that case, a significant increase in activation occurred in several parts of the participatory network, including the prefrontal cortex at the extreme front of the brain, the hippocampus, and a nearby area allied with the hippocampus called the retrosplenial cortex. When less time elapsed between the word and the nonsense letters, people had no awareness of words about themselves. In that case, the inferior temporal cortex became significantly activated, but the participatory network didn't.

These observations show that the human brain can process information about the "self" and ownership both subconsciously and consciously. That being the case, we don't know whether the evolutionary developments discussed here have anything to do with consciousness: in extinct hominins or in modern ones. It could be that the kinds of representations depicted in Figure 10.1 occur in both conscious and subconscious form.

We can't end this book by saying what we're not saying; so, what are we saying? The final section sums it up.

The representational road reconsidered

In Figure 10.2, we present a summary of what we've said in this chapter. We use black type and arrows—placed on a drawing of the human brain—to highlight the representations and forms of memory that (we think) emerged during hominin evolution. These human-specific representations augmented the representations indicated by colored type, which evolved in more-distant ancestors. In Figure 10.3, we illustrate the accretion of novel representations in brains that stand-in for our direct ancestors; and, in Figure 10.4, road signs summarize the innovations that occurred along the evolutionary road to human memory. With these drawings in mind, we can summarize our proposal:

As a series of our direct ancestors faced the problems and opportunities of their time and place, their brains developed specialized representations that

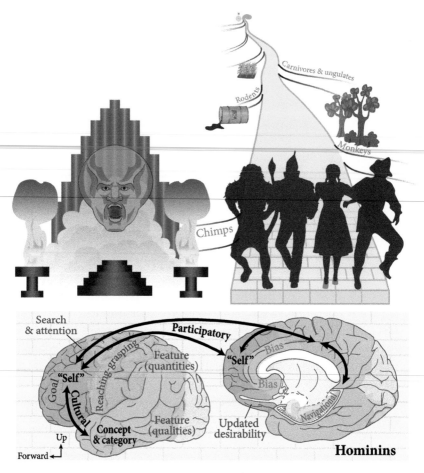

Fig. 10.2 Some hominin augmentations. The representations and forms of memory that emerged during hominin evolution appear in black letters: on a brain presented in the format of Figures 1.2a and 1.2b. Arrows mark parts of the participatory and cultural networks. Older kinds of representations appear in colored letters: with letters and shading color coded as in Figure 10.3. This drawing commemorates the first time that Dorothy and her companions confront the Great and Powerful Oz, a meeting that goes badly. Note that the man behind the curtain projects a representation of his "self" that contrasts dramatically with the humdrum reality of the humbug that he is. In the background, we depict a compass, cornfield, oil can, and forest, which represent additions to Dorothy's support group along the yellow brick road.

helped them survive. Through inheritance across millions of years, these representations—in modified form—made human memory what it is today. Nothing designed, intended, or ordained human memory to have its current character. Indeed, each of the key representations began by doing something quite different from what it does today:

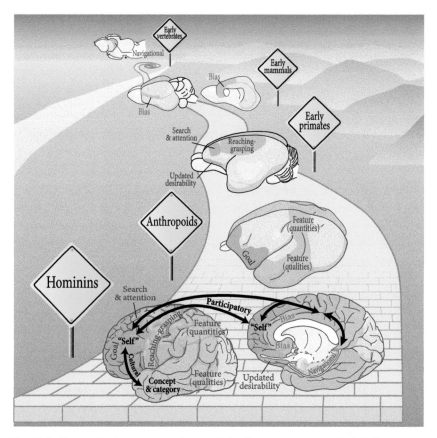

Fig. 10.3 The evolutionary accretion model of human memory. In this figure, we summarize the evolutionary innovations illustrated in Figures 4.4, 5.4, 6.6, 8.2, and 10.2. Each brain along the yellow brick road shows the new representations that emerged in certain ancestors, which provided survival advantages in their time and place. Some of the brains come from modern species, and some are based on fossils. Regardless, each brain stands-in for the brain of a direct, but now-extinct ancestral species: in a direct line of descent from the earliest vertebrates to modern humans.

- Navigational representations, which originally guided vertebrates through their watery world, established the foundation for participatory memory;
- Feature representations, which first guided foraging based on distant sights and sounds, later empowered the generalizations, concepts, and categories of cultural memory;
- Representations of the "self," which initially promoted social cohesion and co-operation, came to underlie the sense of owning knowledge and participating in events: real and imagined.

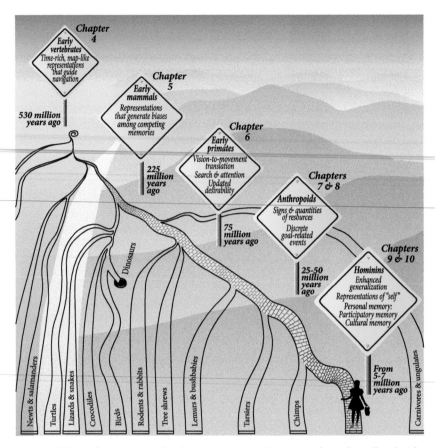

Fig. 10.4 The evolutionary road to human memory. Signposts along the yellow brick road summarize evolutionary innovations along the way.

A lion's lines

In *The Wizard of Oz*, the Cowardly Lion poses a series of questions:

> What makes a king out of a slave? . . .
> What makes the flag on the mast to wave? . . .
> What makes the elephant charge his tusk in the misty mist or the dusky dusk?
> What makes the muskrat guard his musk? . . .
> What makes the Sphinx the Seventh Wonder? . . .
> What makes the dawn come up like thunder?! . . .
> What makes the Hottentot so hot?
> What puts the ape in apricot?
> Whatta they got that I ain't got?

Courage! In the Cowardly Lion's mind, courage is what "they got" that he "ain't got."
But how could a lion know who or what has courage?

The Cowardly Lion can have this knowledge (as well as a mind) because he's a
product of the human brain, which can spin out scenarios that exploit a sense of participation in events, however fictional. In *The Wizard of Oz*, the Lion leads his com-
rades on a courageous raid to free Dorothy from the clutches of the Wicked Witch of
the West, and you remember these events almost as if you experienced them in real
life. What's more, you can read the Lion's mind as he approaches the witch's castle or
takes a cat nap in a poppy field. And the Lion can read minds, too. From the questions
he poses, there seems to be no limit to the kinds of minds he can read: slaves; flags;
elephants; muskrats; the Sphinx; the dawn; a group of people more properly called
the Khoikhoi; and whoever or whatever puts the "ape in apricot." The Cowardly Lion
seems to think that courage gives these "minds" the spark to accomplish great things.
But we know better; courage doesn't make a flag wave or the sun rise. The Cowardly
Lion and his ideas about courage can only arise from a brain that constructs scenarios
of personal and social history, both factual and fictional; and that brain is the human
brain. But human brains are more than that: they are also vertebrate, mammalian,
primate, and anthropoid brains. By appreciating this ancestry in full, we can begin to
understand the evolutionary road to human memory.

References

1. Andrews, P. *An Ape's View of Human Evolution* (Cambridge, UK: Cambridge University Press, 2015).
2. Klein, R. G. *The Human Career: Human Biological and Cultural Origins* (Chicago, IL: University of Chicago Press, 2009).
3. Passingham, R. E. & Smaers, J. B. Is the prefrontal cortex especially enlarged in the human brain? Allometric relations and remapping factors. *Brain, Behavior and Evolution* 84, 156–166 (2014).
4. Dunbar, R. *Human Evolution* (London, UK: Pelican-Penguin, 2014).
5. Krause, J. et al. The derived FOXP2 variant of modern humans was shared with Neandertals. *Current Biology* 17, 1908–1912 (2007).
6. Martinez, I. et al. Human hyoid bones from the middle Pleistocene site of the Sima de los Huesos (Sierra de Atapuerca, Spain). *Journal of Human Evolution* 54, 118–124 (2008).
7. Dediu, D. & Levinson, S. C. On the antiquity of language: the reinterpretation of Neandertal linguistic capacities and its consequences. *Frontiers in Psychology* 4, 397 (2013).
8. Watkins, K. Developmental disorders of speech and language: from genes to brain structure and function. *Progress in Brain Research* 189, 225–238 (2011).
9. Sterelny, K. A Paleolithic reciprocation crisis: symbols, signals, and norms. *Biological Theory*, 9, 1–3 (2014).
10. Passingham, R. E., Bengtsson, S. L., & Lau, H. C. Medial frontal cortex: from self-generated action to reflection on one's own performance. *Trends in Cognitive Sciences* 14, 16–21 (2010).

11. Herculano-Houzel, S. *The Human Advantage: How Our Brain Became Remarkable* (Cambridge, MA: MIT Press, 2016).
12. Barton, R. A. & Venditti, C. Human frontal lobes are not relatively large. *Proceedings of the National Academy of Sciences U.S.A.* **110**, 9001–9006 (2013).
13. Elston, G. N. et al. Specializations of the granular prefrontal cortex of primates: implications for cognitive processing. *Anatomical Record A, Discoveries in Molecular, Cellular and Evolutionary Biology* **288**, 26–35 (2006).
14. Tacikowski, P., Berger, C. C., & Ehrsson, H. H. Dissociating the neural basis of conceptual self-awareness from perceptual awareness and unaware self-processing. *Cerebral Cortex,* 27, 3768–3781 (2017).
15. Graziano, M. S. & Kastner, S. Human consciousness and its relationship to social neuroscience: a novel hypothesis. *Cognitive Neuroscience* **2**, 98–113 (2011).
16. Somerville, L. H., Kelley, W. M., & Heatherton, T. F. Self-esteem modulates medial prefrontal cortical responses to evaluative social feedback. *Cerebral Cortex* 20, 3005–3013 (2010).
17. Behrens, T. E., Hunt, L. T., & Rushworth, M. F. The computation of social behavior. *Science* **324**, 1160–1164 (2009).
18. Harris, L. T. & Fiske, S. T. Social groups that elicit disgust are differentially processed in mPFC. *Social Cognitive and Affective Neuroscience* **2**, 45–51 (2007).
19. Amodio, D. M. & Frith, C. D. Meeting of minds: the medial frontal cortex and social cognition. *Nature Reviews Neuroscience* **7**, 268–277 (2006).
20. Apperly, I. A. Beyond simulation-theory and theory-theory: why social cognitive neuroscience should use its own concepts to study "theory of mind." *Cognition* **107**, 266–283 (2008).
21. Heider, F. & Simmel, M. An experimental study of apparent behavior. *American Journal of Psychology* **57**, 243–259 (1944).
22. Blanke, O., Ortigue, S., Landis, T., & Seeck, M. Stimulating illusory own-body perceptions. *Nature* **419**, 269–270 (2002).
23. Nahab, F. B. et al. The neural processes underlying self-agency. *Cerebral Cortex* **21**, 48–55 (2011).
24. Lau, H. & Rosenthal, D. Empirical support for higher-order theories of conscious awareness. *Trends in Cognitive Sciences* **15**, 365–373 (2011).
25. Schacter, D. L. et al. The future of memory: remembering, imagining, and the brain. *Neuron* **76**, 677–694 (2012).
26. Rumiati, R. I., Carnaghi, A., Improta, E., Diez, A. L., & Silveri, M. C. Social groups have a representation of their own: clues from neuropsychology. *Cognitive Neuroscience* **5**, 85–96 (2014).
27. Hurford, J. R. *The Origins of Grammar: Language in the Light of Evolution* (Oxford, UK: Oxford University Press, 2012).
28. Taubert, J. et al. Amygdala lesions eliminate viewing preferences for faces in rhesus monkeys. *Proceedings of the National Academy of Sciences U.S.A.* **115**, 8043–8048 (2018).

29. LeDoux, J. E. & Pine, D. S. Using neuroscience to help understand fear and anxiety: a two-system framework. *American Journal of Psychiatry* **173**, 1083–1093 (2016).
30. Mithen, S. *The Prehistory of the Mind* (London, UK: Thames and Hudson, 1996).
31. Sneve, M. H. et al. High-expanding regions in primate cortical brain evolution support supramodal cognitive flexibility. *Cerebral Cortex* doi:10.1093/cercor/bhy268 (2018).
32. Dennett, D. C. *Consciousness Explained* (Boston, MA: Little Brown, 1991).
33. Ye, Y. & Gawronski, B. When possessions become part of the self: ownership and implicit self-object linking. *Journal of Experimental Social Psychology* **64**, 72–87 (2016).

Epilogue

Stegosaurus stories—Nuts and neocortex

In *The Shy Stegosaurus of Cricket Creek*, a children's book by Evelyn Sibley Lampman, a stegosaurus named George befriends two children, Joan and Joey, who live in the desert somewhere in the United States. George makes his appearance by stomping on a venomous fellow reptile, thereby saving the kids from certain doom. After millions of years alone, as the last living member of his species, George then overcomes his shyness to chat with them (Fig. E.1). He confesses, for example, that it hurt his feelings when pteranodons taunted him about his tiny brain.

When the kids tell adults about George, it goes badly. A paleontology professor, boarding with the family to hunt for fossils, lays it on the line:

> The last dinosaur lived sixty million years ago . . . The stegosaurus, with whom you claim a speaking acquaintance, lived even longer ago.

All the same, the children know the truth, and they're curious about George. Because he regularly disparages his own intelligence, Joan asks the professor whether dinosaurs of his species were really stupid.

> "Oh, very stupid." The professor nodded emphatically. "The stegosaurus had a brain the size of a walnut. You couldn't have selected a more stupid dinosaur for your playmate if you had tried."

In fact, the brain of a stegosaurus weighed approximately 80 grams (2.8 ounces): much more than a walnut and more like a mid-size tomatillo. Large animals tend to have large brains, so it's usually a good idea to consider brain size and body weight together. A typical stegosaurus tipped the scales at about 3000 kilograms (6500 pounds), 40–50 times that of the average human. Nevertheless, the human brain is about 15 times larger than theirs. So, George did have a pretty small brain, all things considered. The hated pteranodons had bigger brains than most reptiles their size, so perhaps they had a right to tease George about his tiny brain.

Despite these facts, Joey has doubts about the professor's assumption that a small brain implies low intelligence, at least as it applies to his friend George:

> "How can you be so sure it was stupid," asked Joey stubbornly. "Just because its brain was small doesn't prove anything. It might have been different than ours. Maybe it didn't have to be very big . . . "

The Evolutionary Road to Human Memory. Elisabeth A. Murray, Steven P. Wise, Mary K. L. Baldwin, and Kim S. Graham, Oxford University Press (2020). © Oxford University Press.
DOI: 10.1093/oso/9780198828051.001.0001

Fig. E.1 George, a shy stegosaurus, greets Joan and Joey.

Later, Joan tells George that she never thought of him as stupid, and, surprisingly, he agrees:

> "I never did believe it either . . . My brain was small, but I always thought it had great capabilities . . . "

George and the kids have a point. The brain's composition matters at least as much as its size—and probably more. The diminutive stegosaurus brain gave these gargantuan vegans a grandeur of their own.

Although every stegosaurus died long ago, they survive in human memory. A famous *Far Side*® cartoon by Gary Larson, easy to find on the internet,* shows a stegosaurus presenting an audience of fellow dinosaurs with a situation report, called a sit-rep for short. "The picture's pretty bleak gentlemen," he tells them, "the world's climates are changing, the mammals are taking over, and we all have a brain about the size of a walnut."

Again, with the walnut! With so much emphasis placed on the small size of dinosaur brains, the stegosaurus fails to mention the fact that the mammals of his era had brains considerably smaller than a walnut. A tyrannosaurus in the embattled Dinosaur Protection Agency (Fig. E.2), might have presented a better sit-rep: "The mammals have developed a new weapon called neocortex, and they've deployed it; their brains are puny but powerful; and yet, our politicians continue to deny that

---

* Search for the text string "Larson mammals are taking over."

Fig. E.2 A tyrannosaurus delivers a situation report to a governmental committee composed, like many such committees, of dinosaurs.

neocortex exists. We, however, don't have the luxury of ignoring scientific facts. We need to prepare for the Age of Mammals, and we'd better start now."

A news article from the Canadian Broadcasting Corporation in 2014 also mentioned reptilian brains, and it tells a very tall tale indeed:

> The "lizard brain" is a catch-all term for the areas of our brain that developed between 500 million and 150 million years ago and are primarily responsible for instinct, emotion and recording memories . . .
>
> The neocortex, on the other hand, is the area of our brain responsible for reason, language, imagination, abstract thought and consciousness. Scientists say the neocortex has only been around for two or three million years.

The errors in this quotation are many and instructive. First, the idea that we have a "lizard brain" is nonsense. As we explained in Chapter 2, everything in our brain is mammalian. The fact that some parts of our brain evolved before the advent of mammals doesn't mean that we have a "lizard brain." If it did, then the fact that paired appendages—pectoral and pelvic fins—evolved in the first jawed vertebrates would mean that we have fins rather than arms and legs, not to mention the uniquely human configuration of our hands and feet. And, although it's true that parts of our arms and legs have homologs in lizards, we don't have "lizard limbs" either.

Second, the article says that brain areas shared by lizards and humans are "primarily responsible for . . . recording memories." Some such areas—the hippocampus in particular—play an important role in memory, but many of our memories depend on neocortical areas, which evolved in mammals and, as such, are not shared with lizards. The cultural memories discussed in Chapter 9, for example, depend on a region of neocortex that expanded dramatically during hominin evolution and has no counterpart in the brain of any lizard.

Third, the idea that the neocortex emerged just 2–3 million years ago is astoundingly inaccurate. The neocortex first appeared somewhere around 225 million years ago, maybe earlier. Something important happened 2–3 million years ago, but not that. The hominin neocortex expanded dramatically at about that time, and, as we explained in Chapter 10, we think that new representations of the "self" developed when it did. As representations of events, facts, and fictions combined with new representations of the "self," the human experience of participating in events and owning knowledge arose for the first time. Autobiographies, histories, and fabulous fables followed, including a magical tale about the self-discovered self-reliance of Dorothy Gale, a scarecrow who receives a Th.D. (Doctor of Thinkology degree), and an insecure member of the cat family who conquers his fears and earns a medal for "conspicuous bravery against Wicked Witches."

As perhaps its most impressive feat, human memory has the power to connect us with our ancestors: from pioneering primates persisting in the trees to prehistoric people pondering their dreams. And dreams are no small thing. We have a dream that one day science will bring forth a deeper understanding of the evolutionary road to human memory. And someday, it will. Until that day, consider this book our sit-rep for now.

Glossary

Activation* In brain-imaging studies, a statistically significant increase in a measure of brain function (relative to some comparison level)

Activity In this book, a *neuron's* rate of action potentials (also known as pulses, spikes, firing, impulses, and discharges); excludes *activation*

Allocortex Three-layered cortex; contrasts with *neocortex*

Amniote An animal in the lineage that includes *reptiles*, birds, and mammals

Amphioxus A *deuterostome* species, also known as the lancelet

Amygdala A part of the *telencephalon*

Analogy 1. In biology, a *trait* that has a common function in two or more species; 2. In cognitive psychology, a similarity in relations among items or propositions

Anthropoid An animal in the lineage that includes monkeys, *apes*, and humans

Ape An animal in the lineage that includes gibbons, gorillas, orangutans, chimpanzees, and *bonobos*

Association A linkage between unique *representations* in the brain

Australopithecine One of several extinct *hominin* species

Basal ganglia A part of the *telencephalon*, thought by some scientists (but not by us) to be the "habit center" of the brain

Behaviorism A school of psychology emphasizing *conditioned reflexes* as an account for human and animal cognition

Bilateral In both hemispheres of the brain or both sides of the body

Bonobo An ape species closely related to chimpanzees

Cerebral cortex The brain's outer covering and the largest part of the human brain; includes *allocortex* and *neocortex*; often called simply "the cortex"

CHAP An acronym for the chimpanzee–human ancestral population, the last common ancestor of humans and chimpanzees

Cognition Knowledge and perceptions; or the process of generating them

Cognitive domain A kind of knowledge: about society, the physical world, or tools and technologies, for example

Cognitive map See *navigational memory*

Conditioned reflex An umbrella term for a *Pavlovian memory*, an *instrumental memory*, or a *habit*

---

* Words in italic type are included in this glossary.

Context A set of sensations, situations, or circumstances, but sometimes just a single stimulus; also known as a behavioral context

Convergent evolution/convergence Independent evolutionary developments that result in similar *traits*

Core neocortex *Neocortical* areas most remote from *allocortical* areas; contrasts with *ring neocortex*

Cortical Pertaining to the cerebral cortex

Cortical hub A cortical area that draws on and integrates representations from other cortical areas

Cultural memory A *personal memory* of a fact, a fiction, a concept, a category, or a generalization; also known as a semantic memory

Declarative memory See *personal memory*

Dementia A condition characterized by loss of, or inability to access, previously stored knowledge; a group of brain disorders that impair memory

Deuterostome A group of animals that includes starfish, sea urchins, tunicates, lancelets (also known as *Amphioxus*), and vertebrates; contrasts with *protostome*

Devaluation In this book, the process of decreasing the *desirability* of an *outcome*, often by *selective satiation*

Domain See *cognitive domain*

Encephalization quotient The ratio of actual brain size to the brain size predicted for a species with its body weight (for a particular group of animals)

Entorhinal cortex A part of the *cerebral cortex* near the *amygdala* and the *hippocampus*

Episodic memory See *participatory memory*

EQ See *encephalization quotient*

Explicit memory See *personal memory*

Extinction 1. In biology, the demise of a species or larger group of organisms; 2. In psychology, the suppression of a *conditioned reflex* in the absence of a predicted *outcome*

Feature ambiguity The cause of difficulty in distinguishing items because of *feature overlap*

Feature conjunction A combination of features incorporated into a unique neural representation

Feature overlap Sensory properties in common among different items

Fovea A specialized part of the retina with a high density of color detectors and high acuity; the source of central (as opposed to peripheral) vision

Frontal cortex/frontal lobe A part of the *cerebral cortex* toward the front of the brain

Frontal eye field A part of the *cerebral cortex* in the middle of the *frontal lobe*

Habit 1. In psychology, either a behavior based on a stimulus–response *association* without reference to a predicted *outcome* or a behavior performed while attending to something else; 2. In biology, a genetically encoded *trait*, such as the fossorial habit of certain rodents

Haplorhine An animal in the lineage that includes *anthropoids* and tarsiers

Hippocampus/hippocampal A large part of the *allocortex* in *amniotes*, along with *homologous* areas in other vertebrates

Hominin A branch of the *CHAPs* descendants that includes modern and extinct humans, as well as *Australopithecines*

Homology/homolog/homologous A *trait* inherited from a common ancestor in two or more species

Hub See *cortical hub*

Inferior temporal cortex A set of visual areas in the *temporal lobe* that specializes in representing visual *feature conjunctions* that may include colors, shapes, and surface textures; also known as inferotemporal cortex

Instrumental memory The memory of an *association* between an action and an *outcome*, sometimes including an associated sensation

Lancelet A *deuterostome* closely related to *vertebrates*; also known as *Amphioxus*

Lineage A group of animals that descended from a common ancestor

Long-term memory A memory stored for hours, days, weeks, or years

Medial temporal lobe A set of four cortical areas believed by some scientists (but not by us) to underlie all human and animal memory

Memory Stored information acquired during an individual's lifetime (or the capacity to store such information); excludes instincts and reflexes

Memory system A set of brain structures that specializes in the storage and processing of a particular kind of neural *representation*

Navigational memory A time-rich, map-like *representation* that guide journeys through an animal's environment (and contributes to other cognitive functions)

Neocortex Six-layered cortex, specific to mammals; contrasts with *allocortex*

Neuron An information-processing cell in the brain or elsewhere in the nervous system

New World primate An animal in the lineage that includes all South American monkeys but excludes *Old World primates*

Old World primate An animal in the lineage that includes Old World (African and Asian) monkeys, apes, and humans

Optogenetic A method for activating or blocking connections between *neurons*, turned on and off like a light switch

Orbital frontal cortex A part of the *prefrontal cortex*; also known as the orbitofrontal cortex and the orbital prefrontal cortex

Outcome An event that follows a sensation or an action; examples include foods, fluids, and electric shocks; also known as an unconditioned stimulus

Outcome-based behavior An action based on an *association* between a sensation and an *outcome* or between an action and an *outcome*; also known as a goal-directed behavior

Parahippocampal cortex A part of the *neocortex* near the *amygdala* and the *hippocampus*

Parietal cortex/parietal lobe A part of the *neocortex* toward the top, back of the brain

Participatory memory A *personal memory* of an event, including the perception of participating in or experiencing the event; also known as episodic memory

Pavlovian conditioning The process of establishing a *Pavlovian memory*; also known as training

Pavlovian memory A *representation* that records an *association* between a sensation and an *outcome* (or, equivalently, between a conditioned stimulus and an unconditioned stimulus)

Perceptual learning An improvement, with practice, in the ability to distinguish similar sensations, such as slightly different shades of a color

Perirhinal cortex A part of the *cerebral cortex* near the *amygdala* and the *hippocampus*; part of the *ring neocortex*

Personal memory A memory accompanied by the sense of participating in a remembered event or owning knowledge; also known as an explicit memory or a declarative memory

Prefrontal cortex A part of the *neocortex* toward the front of the *frontal lobe*

Premotor cortex A part of the *frontal lobe* between the *prefrontal cortex* and the primary motor cortex

Prosimian An umbrella term for small, nocturnal primates, including lemurs, bushbabies, and tarsiers

Protostome A group of animals that includes most invertebrate species, including insects, mollusks, and crustaceans; contrasts with *deuterostome*

Representation The product of an interconnected network of *neurons*, including actively processed and stored information

Reptile An *amniote* other than a bird or a mammal, including extinct species such as dinosaurs and pteranadons

Ring neocortex A part of the *neocortex* adjacent to *allocortex*; contrasts with *core neocortex*

Selective satiation The *devaluation* a specific kind of food or fluid after consuming it in large quantities

Semantic dementia A progressive brain disease that causes the decline and loss of *cultural memories*

Short-term memory A temporary form of memory that usually persists for a few seconds or minutes; see also *working memory*

Species-specific trait A *trait* of a species that other species either lack or have in a different form; also known as a species-typical *trait*

Superior temporal cortex A set of auditory areas in the *temporal lobe* that specializes in representing acoustic *feature conjunctions* that may include pitch, tone, melody, and prosody

Telencephalon The endbrain; the front part of *vertebrate* brains

Temporal cortex/temporal lobe A part of the *neocortex* under the temple; the lower back part of the *cerebral cortex*; includes the *inferior temporal cortex* and the *superior temporal cortex*

Trait A characteristic of a species, individual, or group of species

Trichromatic vision Vision based on three varieties of color detectors; also known as full-color vision

Tunicate A *deuterostome* closely related to *vertebrates*

Ventral prefrontal cortex A part of the *anthropoid prefrontal cortex*

Ventral premotor cortex A part of the primate *premotor cortex*

Vertebrates A group of *deuterostomes* that includes fishes, amphibians, *reptiles*, mammals, and birds

Working memory A temporary form of memory that involves both maintaining and manipulating information

Index